MY FIRST Piano Sheet Music

Easy, Fun-to-Play Popular Songs for Kids

Emily Norris

Illustrated by Malgorzata Detner

T0020722

Z Kids • New York

CONTENTS

Introduction

Hi,

I'm G-sharp the Giraffe! If you were on our first journey together in *My First Piano Lessons*, you already know me. If not, welcome! I'm so excited to meet you.

This book has 40 of my favorite songs. They are easy to play and perfect for kids new to the piano, just like you!

The first songs in this book are very easy and use only your right hand. As our journey continues, the songs get a little harder, and the ones at the very end will use both hands. But don't worry, each song will have hand positions and a few labeled and numbered notes to help guide you. You have all the knowledge you need to play these songs beautifully.

Some songs might sound a little different than you remember them. That's because I changed a few notes to make it easier for beginners to play. But you will still recognize a lot of the melodies. See if you can hear the difference in some of the songs!

You can play the songs in order or you can flip through the book and pick out your favorites to learn first. To keep things fun as you learn to play these songs, I'll pop in now and again with a cool fact or two about music and the piano!

Remember that:

Journeys can be tiring.
So take breaks!

Journeys can be hard.
It's OK to make mistakes!

Journeys are exciting.
So let's have fun!

Merrily We Roll Along

1 2 3 4 5
C D E F G
3 2 3 5
E D C D E E E D D D E G G
Mer - ri - ly we roll a - long, roll a - long, roll a - long.
3 2
E D C D E E E D D E D C
Mer - ri - ly we roll a - long, o'er the deep blue sea.

FUN FACT

The word "music" comes from the Greek word mousikē, meaning "art of the Muses." In Greek mythology, Muses were goddesses who represented poetry, art, science, and, yes, music!

Rain, Rain, Go Away

1 2 3 4 5
C D E F G

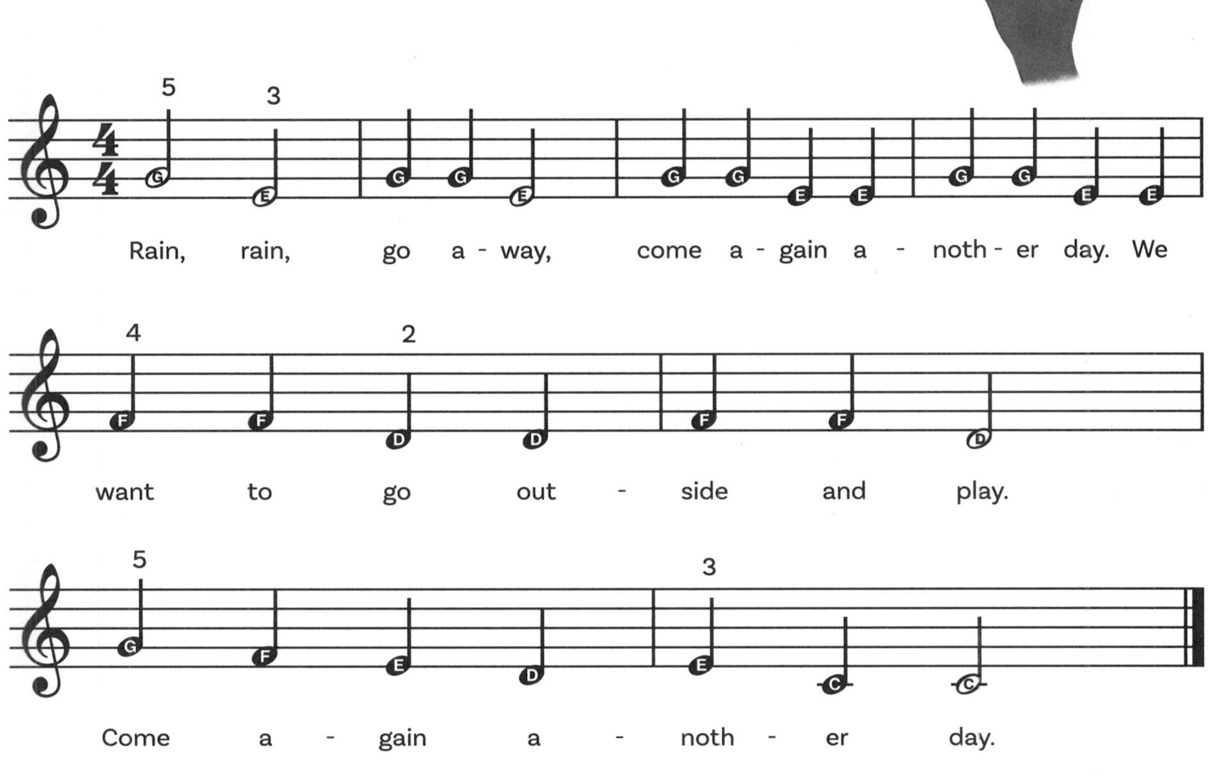

5
3
G E G G E G G E E G G E E
Rain, rain, go a - way, come a - gain a - noth - er day. We
4
2
F F D D F F D
want to go out - side and play.
5
3
G F E D E C C
Come a - gain a - noth - er day.

On the Bridge of Avignon

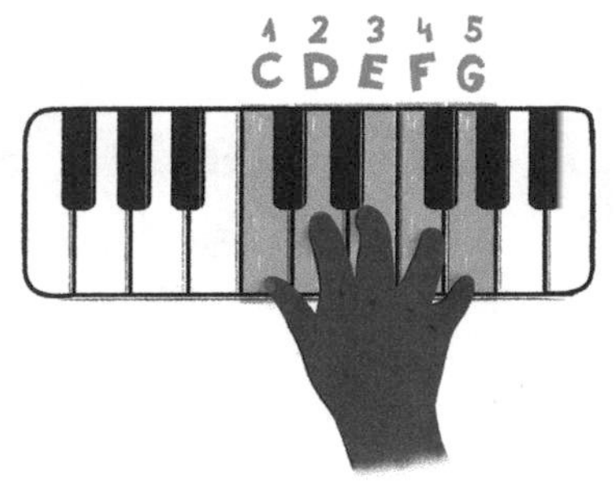
1 2 3 4 5
C D E F G

1
C C C C D D D
On the bridge of Av - i - gnon,
3 5 1
E F G C C C D D
they are danc - ing, they are danc - ing.
C C C C D D D
On the bridge of Av - i - gnon,
5 1 2
E F G C D D C
they are danc - ing all a - round.

I Saw Three Ships

1 2 3 4 5
C D E F G
1 5 3 4
C C C D E G E D F
I saw three ships come sail - ing in, on
Oh let us all re - joice and sing, on
3 3 5
E C C E D D G C
Christ - mas Day, on Christ - mas Day. I
Christ - mas Day, on Christ - mas Day. Oh,
5 3 4
C C D E G E D F
saw three ships come sail - ing in, on
let us all re - joice and sing, on
3
E C C D E D C
Christ - mas day in the morn - ing.
Christ - mas day in the morn - ing.

Dreidel Song

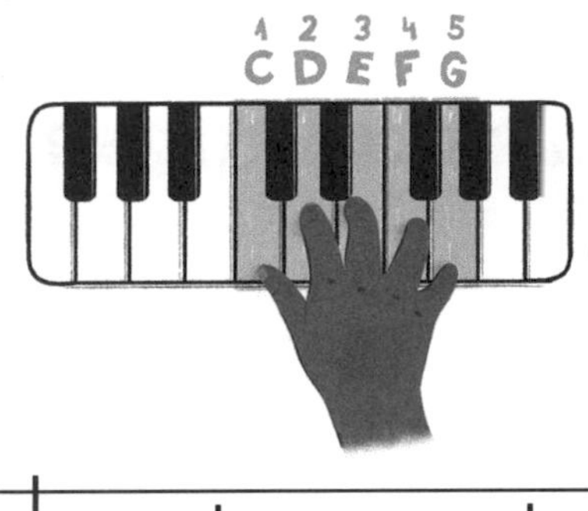
1 2 3 4 5
C D E F G

3 5 3
E G E G E G E E
I have a lit - tle drei - del, I

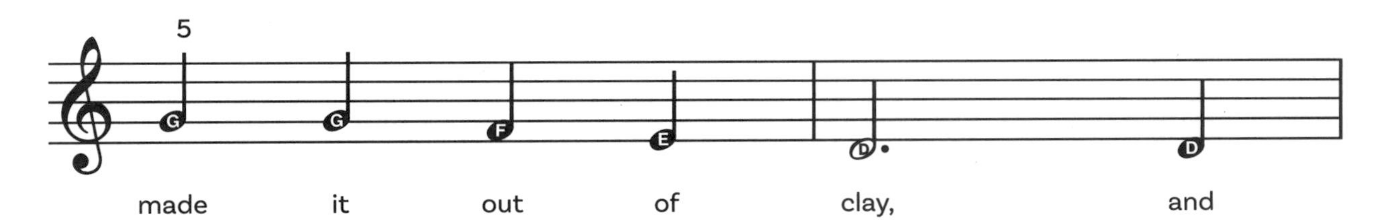
5
G G F E D D
made it out of clay, and

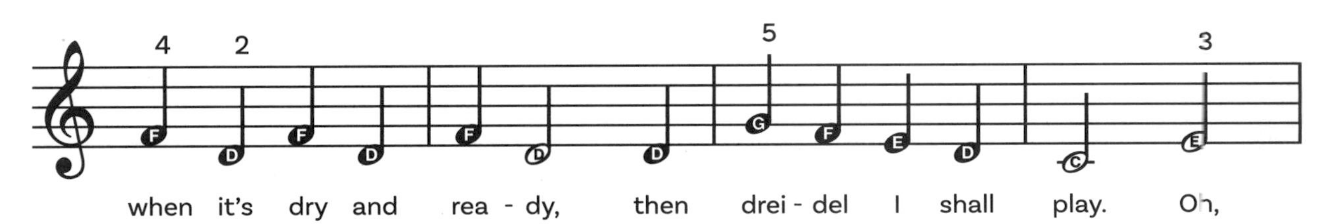
4 2 5 3
F D F D F D D G F E D C E
when it's dry and rea - dy, then drei - del I shall play. Oh,

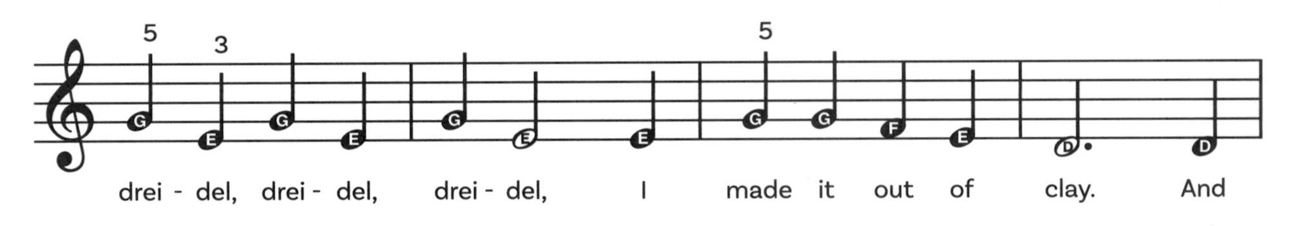
5 3 5
G E G E G E E G G F E D D
drei - del, drei - del, drei - del, I made it out of clay. And

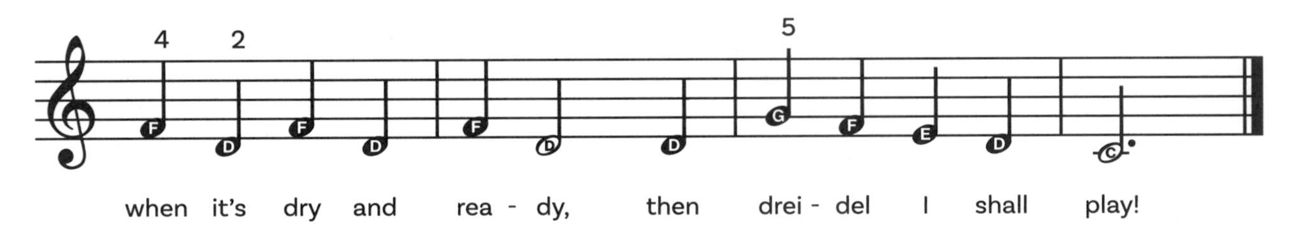
4 2 5
F D F D F D D G F E D C
when it's dry and rea - dy, then drei - del I shall play!

When the Saints Go Marching In

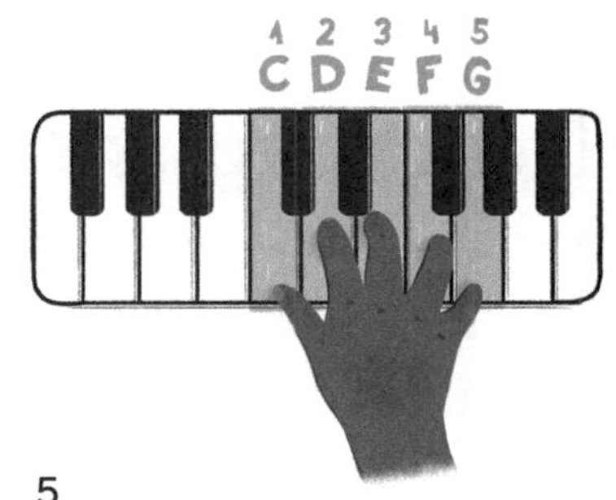
1 2 3 4 5
C D E F G

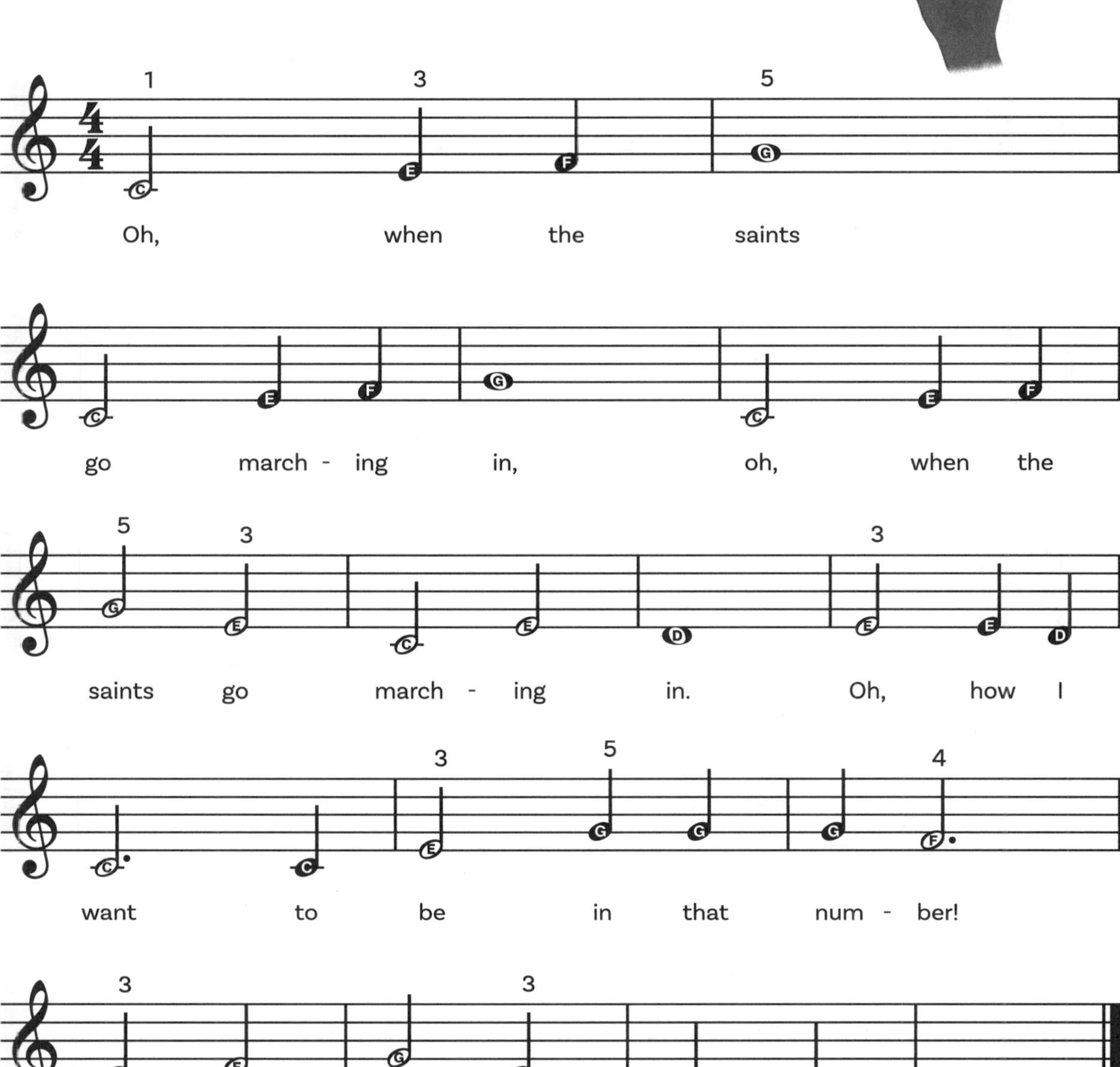
1
3
5
C
E
F
G
Oh, when the saints
C
E
F
G
C
E
F
go march - ing in, oh, when the
5
3
3
G
E
C
E
D
E
E
D
saints go march - ing in. Oh, how I
3
5
4
C
C
E
G
G
G
F
want to be in that num - ber!
3
3
E
F
G
E
D
D
C
When the saints go march - ing in.

Down by the Station

1 2 3 4 5
C D E F G

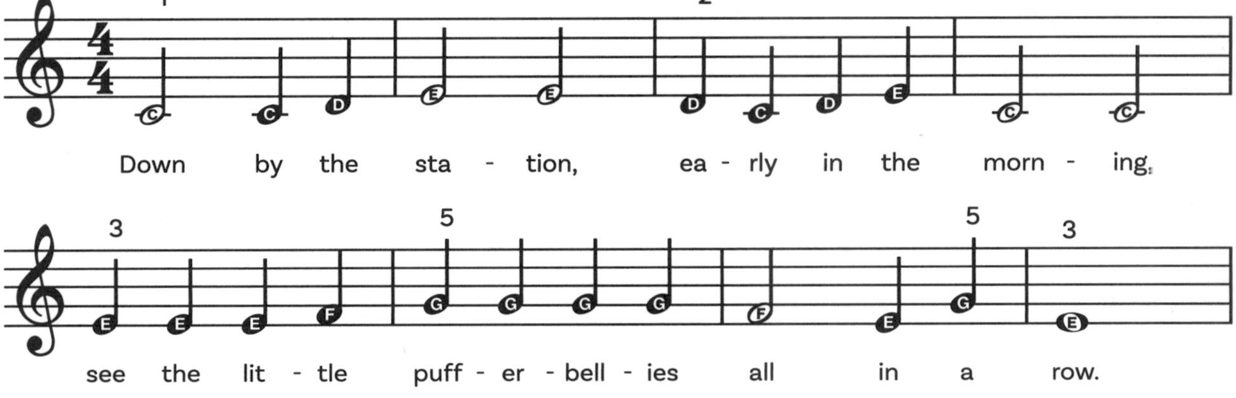
1 2
Down by the sta - tion, ea - rly in the morn - ing.
3 5 5 3
see the lit - tle puff - er - bell - ies all in a row.

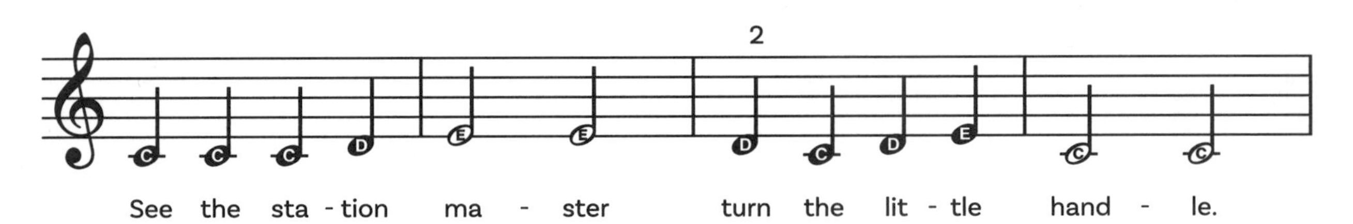
2
See the sta - tion ma - ster turn the lit - tle hand - le.

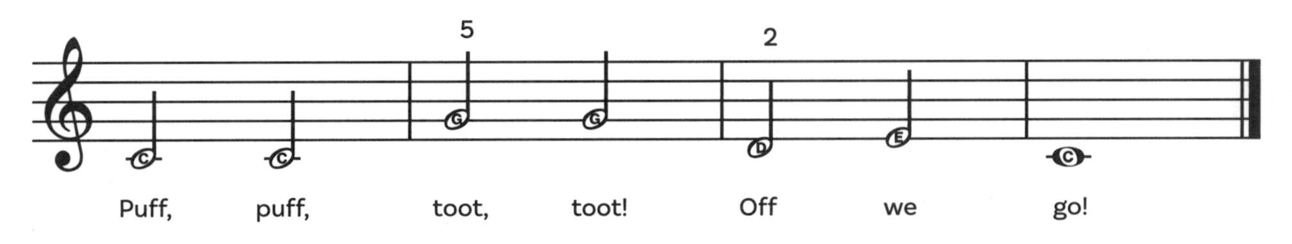
5 2
Puff, puff, toot, toot! Off we go!

Pat-a-Cake

1 2 3 4 5
C D E F G

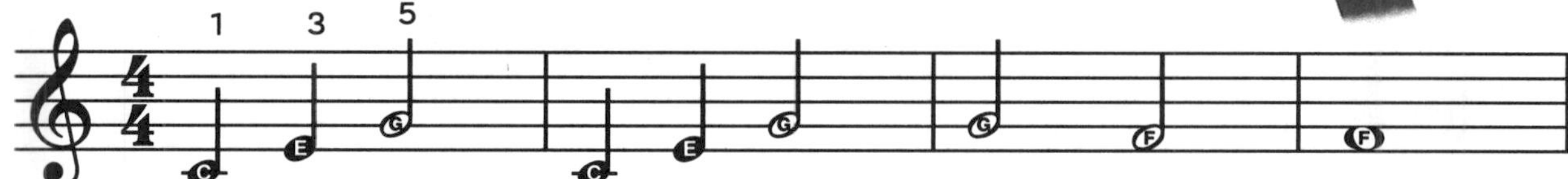
1 3 5
Pat - a - cake, pat - a - cake, bak - er's man.

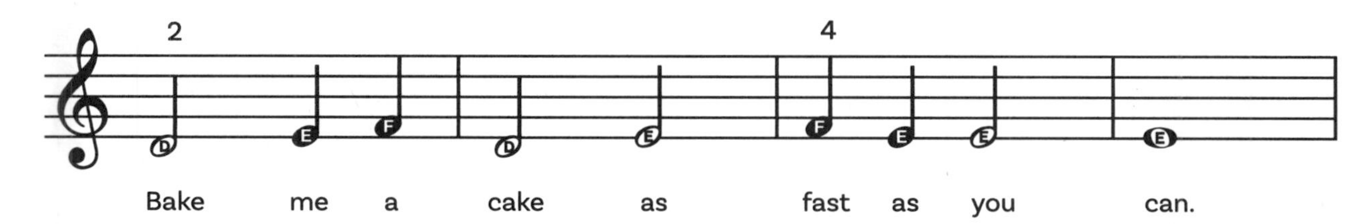
2 4
Bake me a cake as fast as you can.

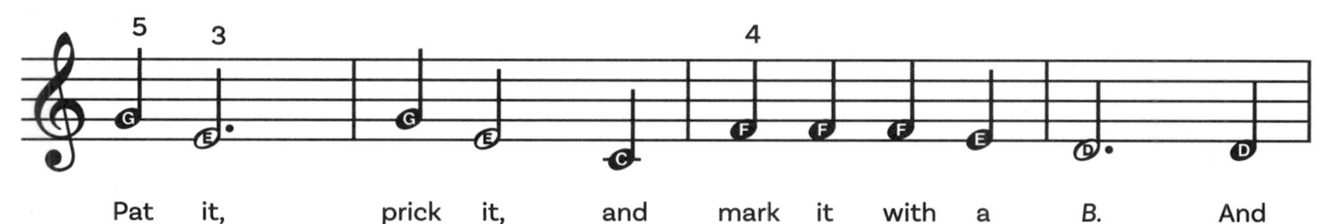
5 3 4
Pat it, prick it, and mark it with a *B*. And

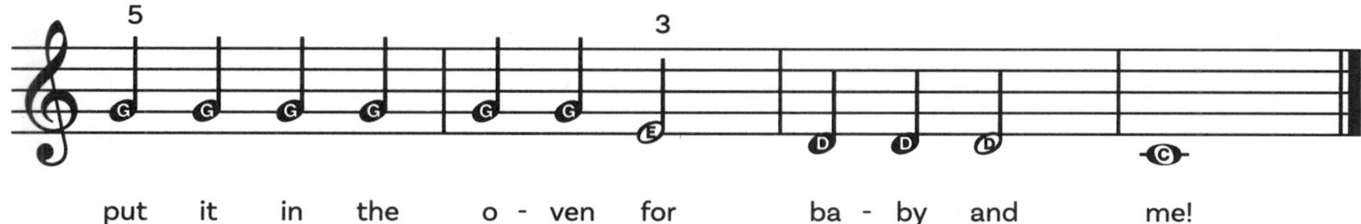
5 3
put it in the o - ven for ba - by and me!

FUN FACT

The first collection of nursery rhymes was published in 1744. But nursery rhymes are older than that—some even date back to the 1100s!

Oats, Peas, Beans, and Barley Grow

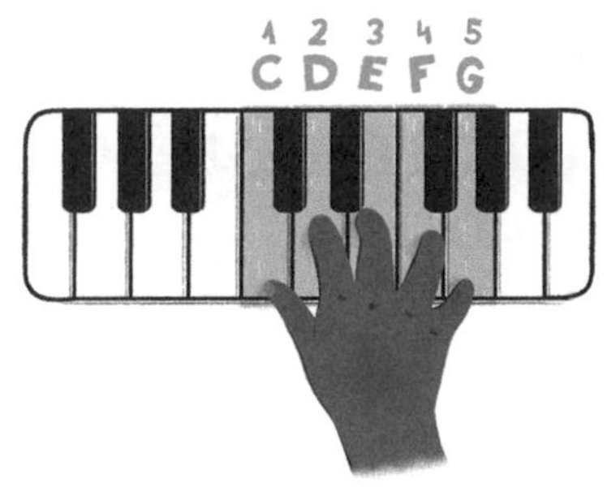
1 2 3 4 5
C D E F G

3
Oats, peas, beans, and bar - ley
4
2
grow. Oats, peas, beans, and bar - ley
5
3
grow. Do you or I or a - ny - one know how
4
oats, peas, beans, and bar - ley grow?

Little Bo-Peep

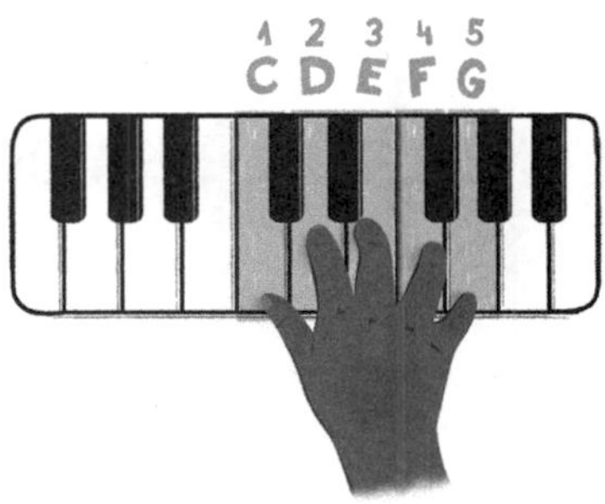
1 2 3 4 5
C D E F G

1
C C C C C D D D D
Lit - tle Bo - Peep has lost her sheep, and
3 5 3
E G G E E D
can't tell where to find them.
5 3 2
G E E E F D D D
Leave them a - lone and they'll come home
3 3
E F G E D D C
wagg - ing their tails be - hind them.

Hush, Little Baby

1 2 3 4 5
C D E F G

3
E E E E F E D D
Hush, lit - tle ba - by, don't say a

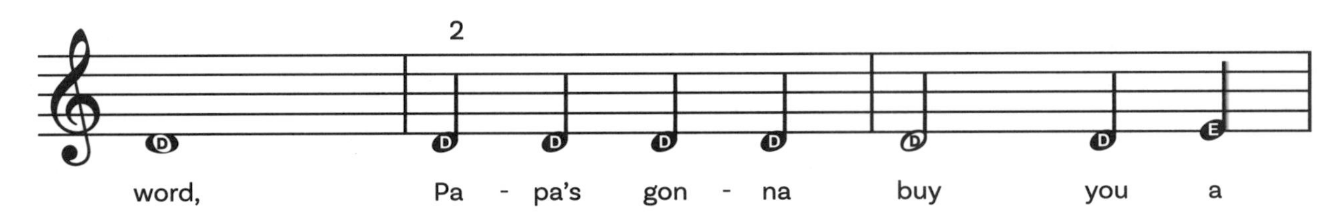
2
D D D D D D D E
word, Pa - pa's gon - na buy you a

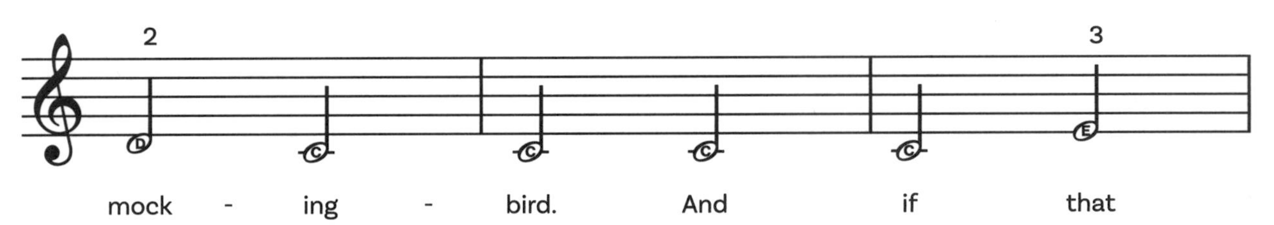
2 3
D C C C C E
mock - ing - bird. And if that

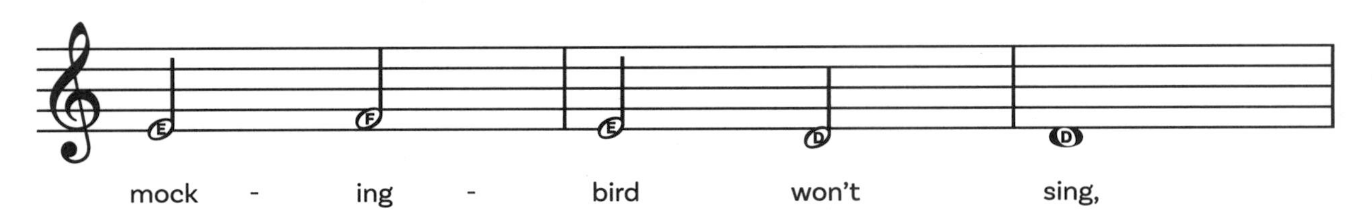
E F E D D
mock - ing - bird won't sing,

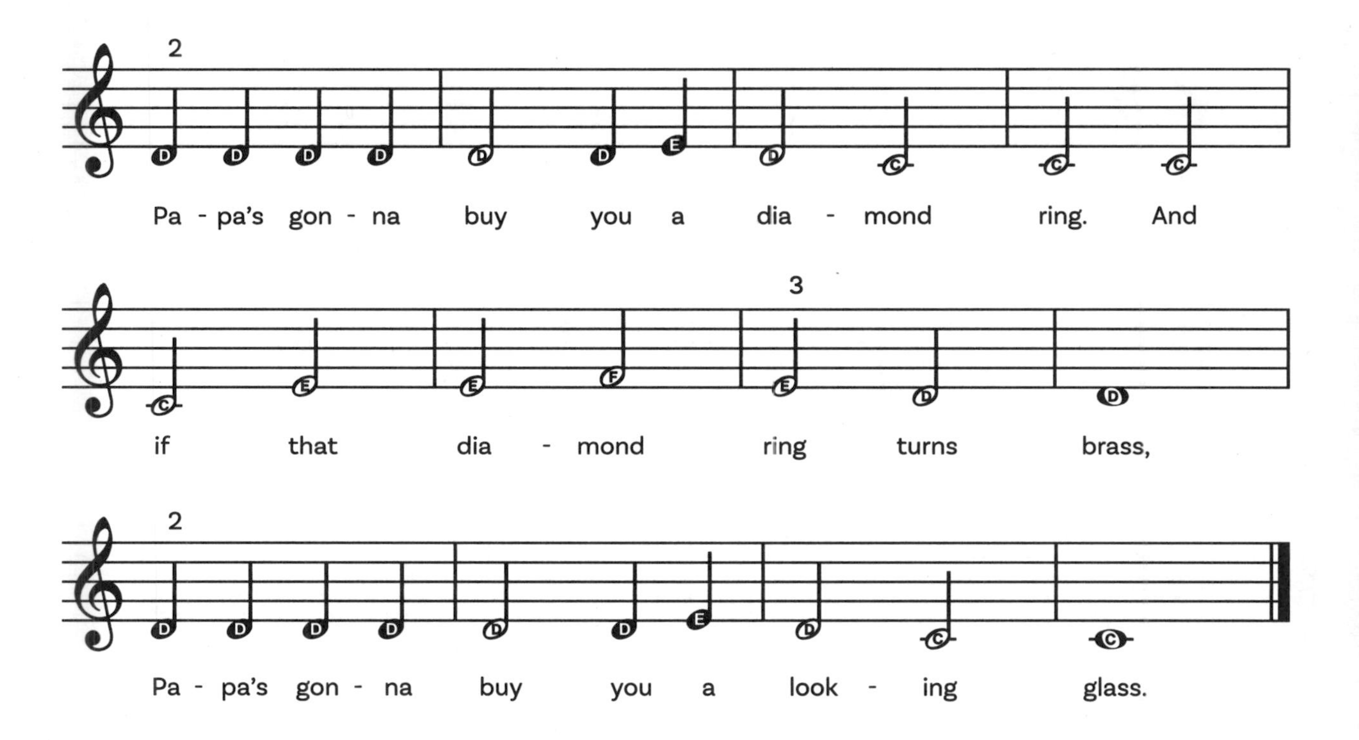
2
D D D D D D E D C C C
Pa - pa's gon - na buy you a dia - mond ring. And
3
C E E F E D D
if that dia - mond ring turns brass,
2
D D D D D D E D C C
Pa - pa's gon - na buy you a look - ing glass.

FUN FACT

Can you guess the price of the most expensive grand piano ever? $3.22 million! It's made of crystal and was used at the 2008 Olympics.

Give My Regards to Broadway

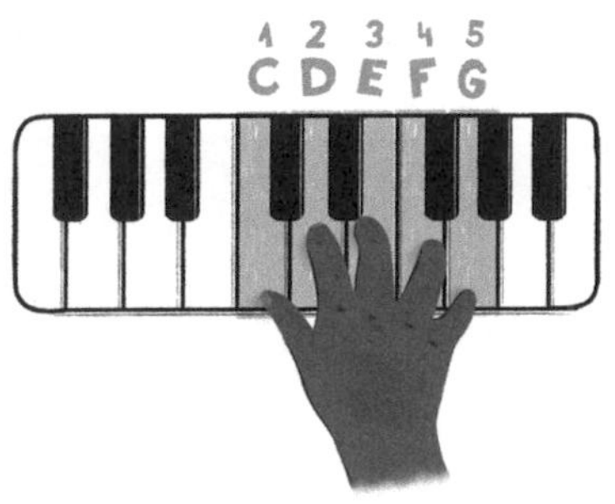
1 2 3 4 5
C D E F G

1
Give my re - gards to Broad - way, re-
3
mem - ber me to Her - ald Square.
1 2
Tell all the gang at For - ty
5 3 4 2
Se - cond Street that I will soon be there.

1
C D E F G F E C
Whis - per of how I'm yearn - ing to
2
D D D D E E C D D C
ming - le with the old - time throng. Give my re-
4
3
F G E F G F G F
gards to old Broad - way, and say that
3
4
G E F G F
I'll be there, e'er long.

Looby Loo

1 2 3 4 5
C D E F G
1
3
5
Here we go loo - by loo,
3
2
here we go loo - by light,
3
here we go loo - by loo,
all on a Sat - ur - day night.
You put your right hand in.
3
You take your right hand out.

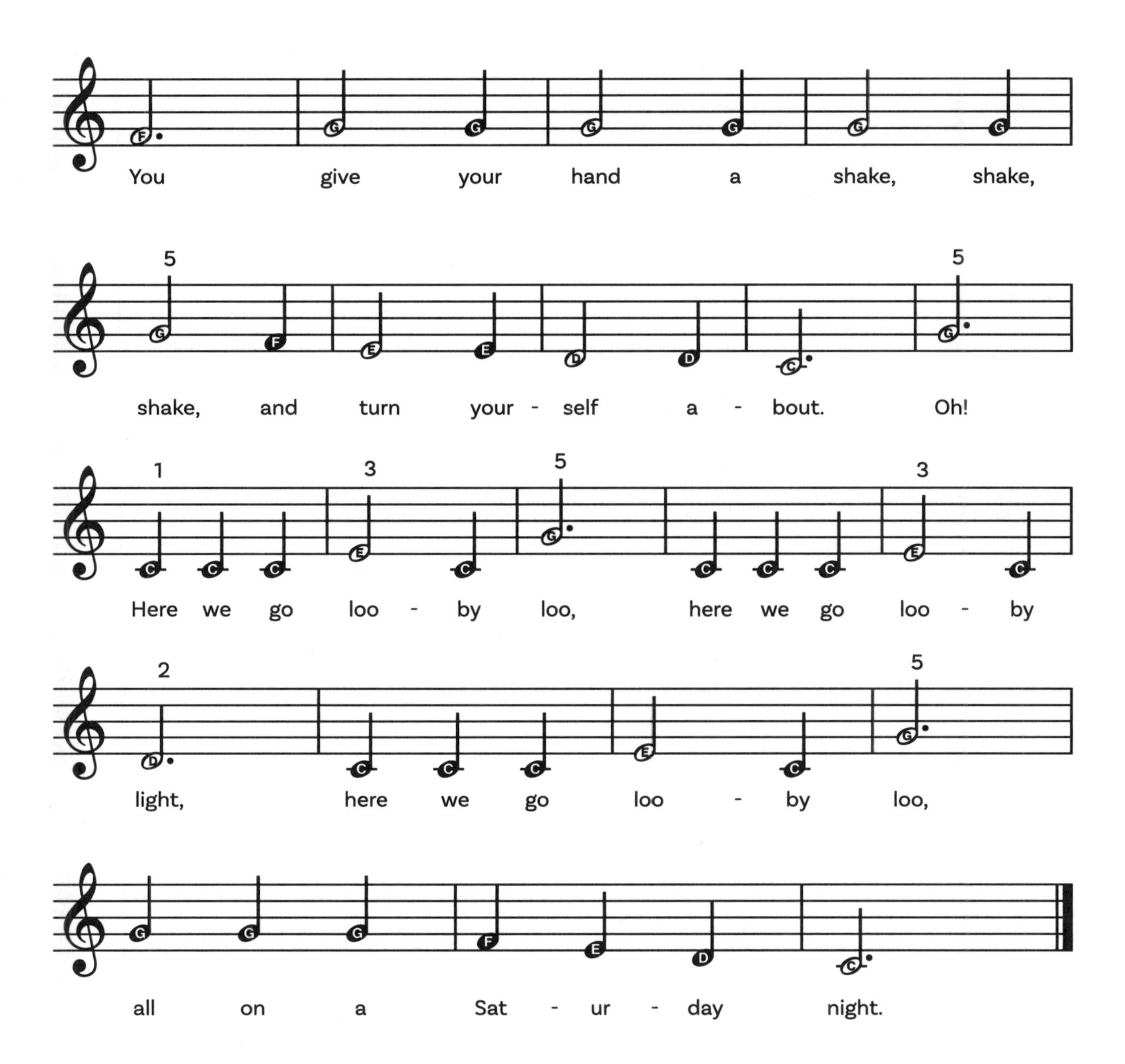
You give your hand a shake, shake,
5 5
shake, and turn your - self a - bout. Oh!
1 3 5 3
Here we go loo - by loo, here we go loo - by
2 5
light, here we go loo - by loo,
all on a Sat - ur - day night.

Stars and Stripes Forever

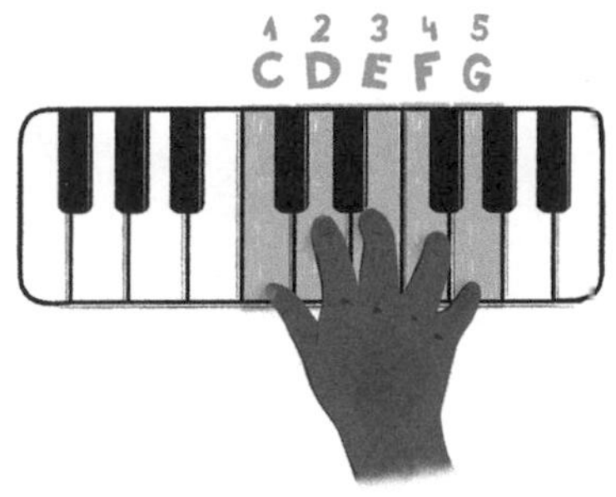
1 2 3 4 5
C D E F G

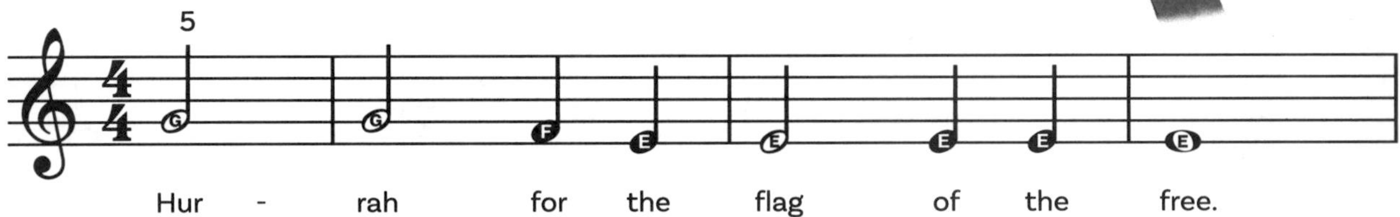
5
G G F E E E E E
Hur - rah for the flag of the free.

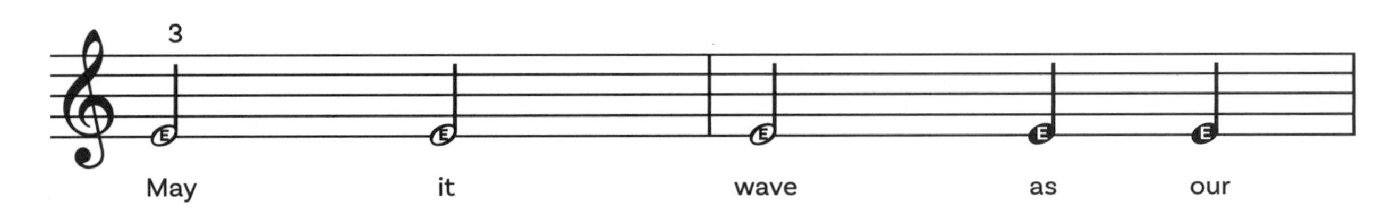
3
E E E E E
May it wave as our

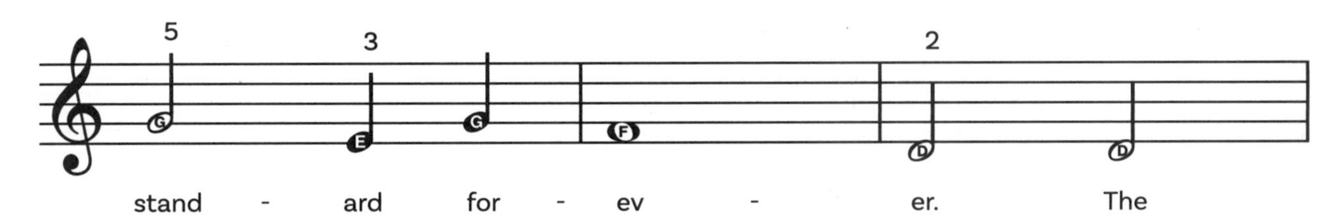
5 3 2
G E G F D D
stand - ard for - ev - er. The

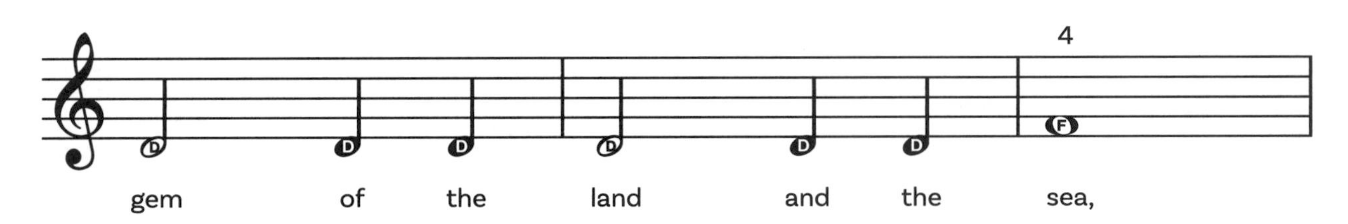
4
D D D D D D F
gem of the land and the sea,

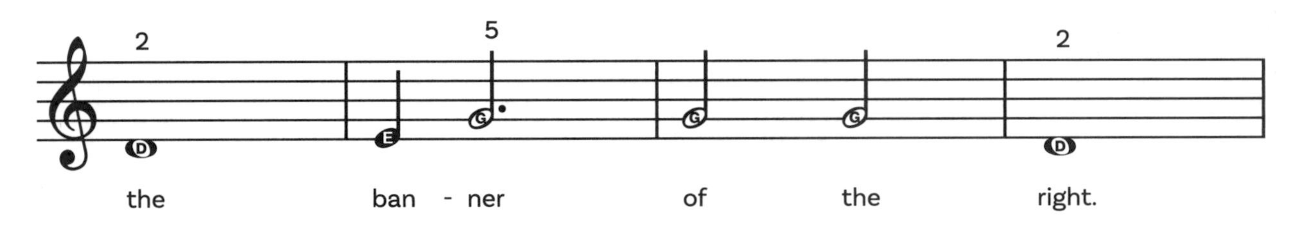
2 5 2
D E G G G D
the ban - ner of the right.

5
G G F E E E E E
Let des - pots re - mem - ber the day
3
E E E D E
when our fa - thers with
F E D C D C C
might - y en - dea - vor, pro-
5
C C C C C C G C C D
claim'd as they march'd to the fray, that by their
3
E C C D E C D E D C
might, and by their right, it waves for - e - ver!

He's Got the Whole World in His Hands

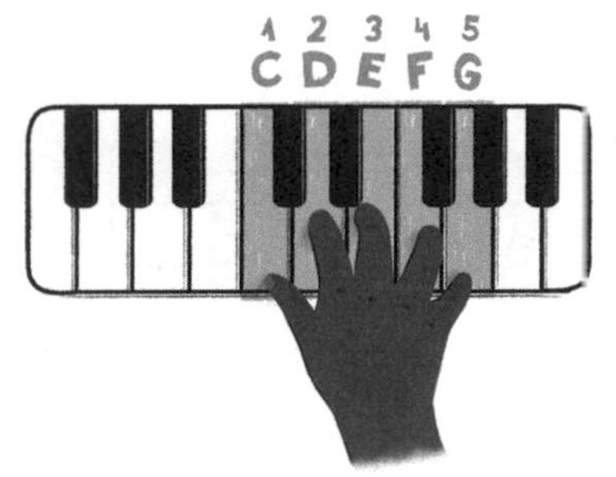

1 2 3 4 5
C D E F G

5 3 5
He's got the whole world
5 3 4 2
in his hands, he's got the
4 2 5
whole world in his hands, he's got the
5 3
whole world in his hands, he's got the
5 4 2
whole world in his hands. He's got my

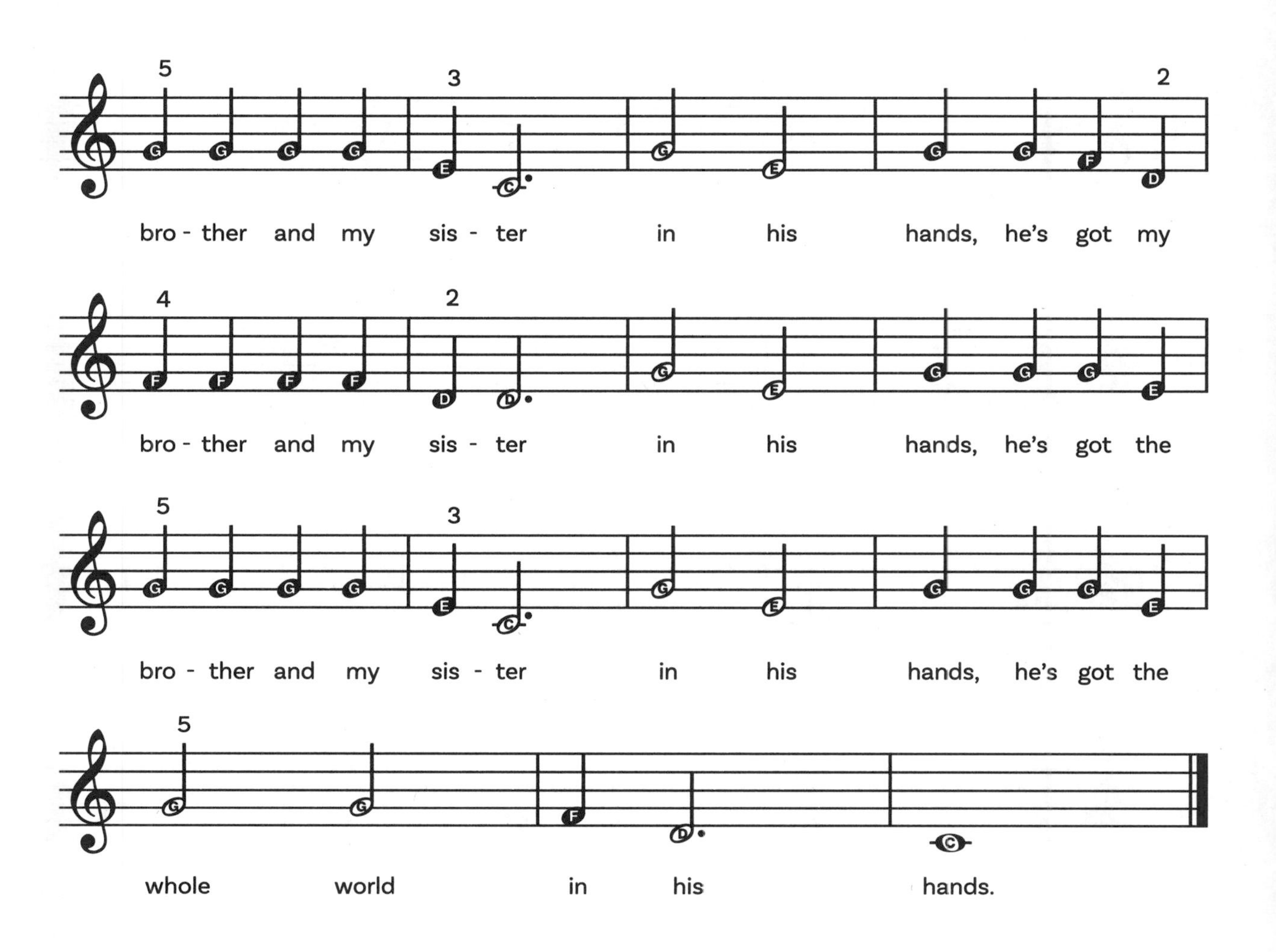
5
3
2
G G G G E C G E G G F D
bro - ther and my sis - ter in his hands, he's got my
4
2
F F F F D D G E G G G E
bro - ther and my sis - ter in his hands, he's got the
5
3
G G G G E C G E G G G E
bro - ther and my sis - ter in his hands, he's got the
5
G G F D C
whole world in his hands.

FUN FACT

Chris Hadfield, a Canadian astronaut, recorded 11 songs for his album while orbiting in space. Music can literally be out of this world!

John Jacob Jingleheimer Schmidt

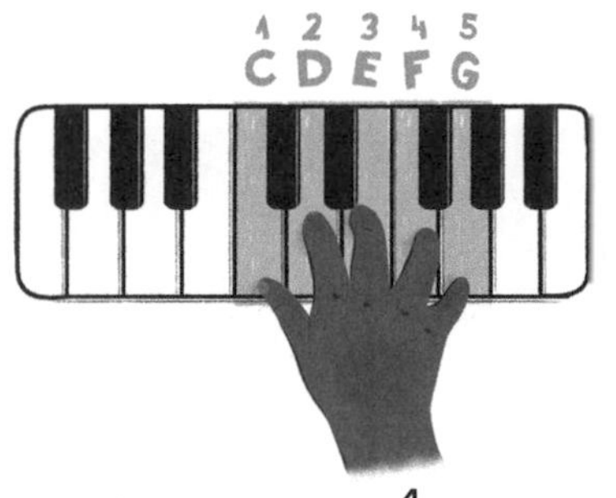
1 2 3 4 5
C D E F G

3
4
4
E
D
C
C
F
F
E
F
f John Ja - cob Jin - gle - hei - mer Schmidt,
5
2
F
F
F
G
D
E
C
his name is my name, too. When -
4
C
E
E
E
E
D
C
F
F
F
F
F
F
e - ver we go out, the peo - ple al - ways shout, "There goes
5
G
G
G
G
F
E
D
John Ja - cob Jin - gle - hei - mer
5
C
G
G
G
G
F
E
D
Schmidt!" Da da da da da da da!

Hokey Pokey

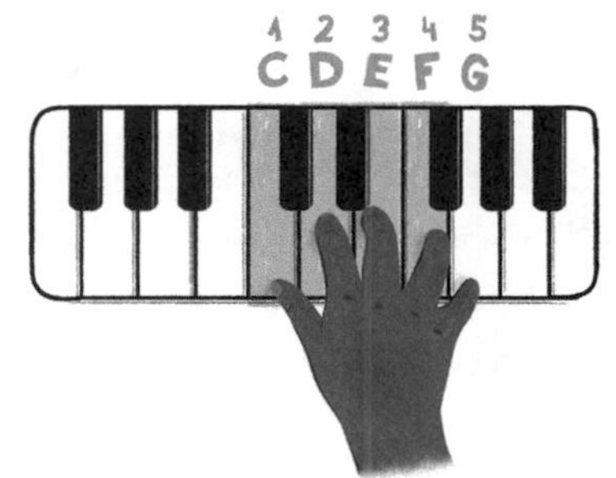
1 2 3 4 5
C D E F G

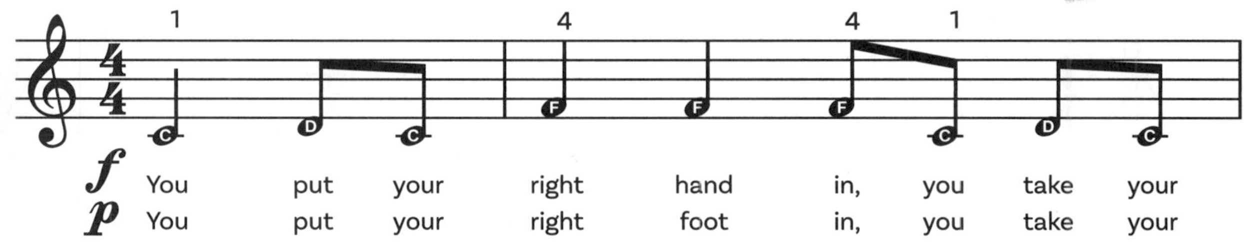
You put your right hand in, you take your
You put your right foot in, you take your

right hand out, you put your right hand in, and you
right foot out, you put your right foot in, and you

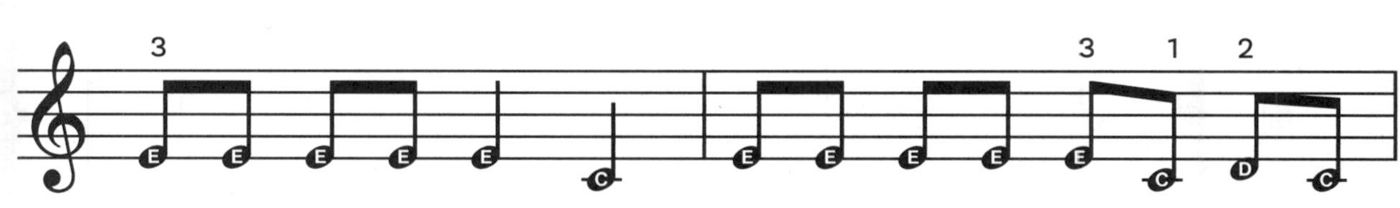
shake it all a - bout. You do the ho - key po - key and you
shake it all a - bout. You do the ho - key po - key and you

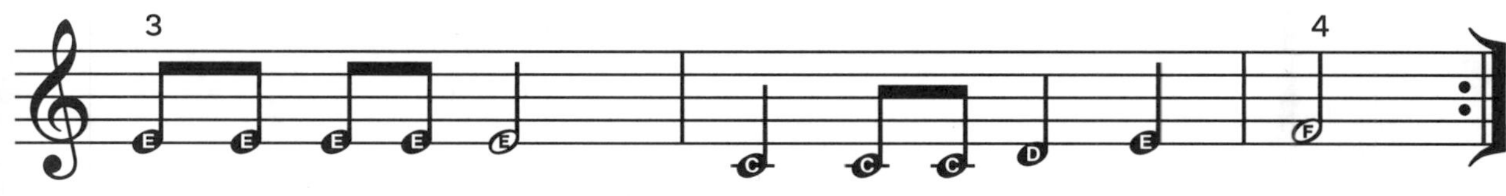
turn your - self a - round, that's what it's all a - bout!
turn your - self a - round, that's what it's all a - bout!

Mulberry Bush

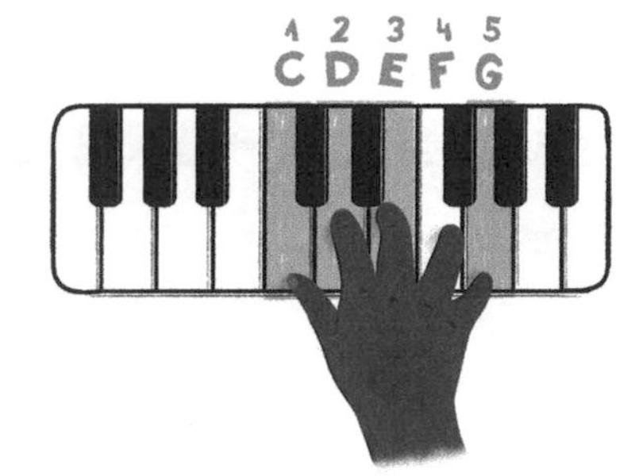

1 2 3 4 5
C D E F G

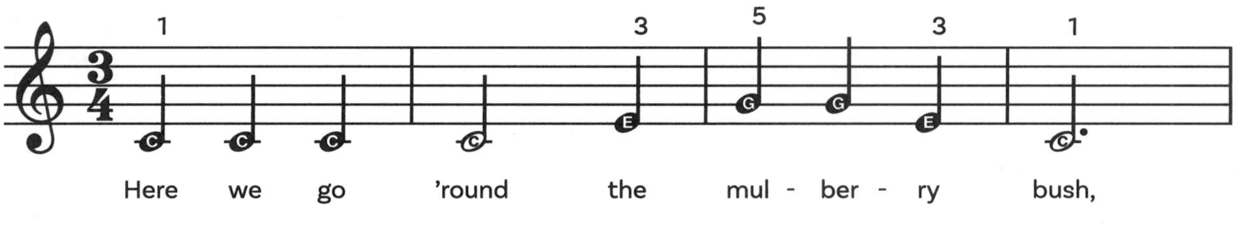

1
3
5
3
1
C C C C E G G E C
Here we go 'round the mul - ber - ry bush,

2
D D D D E E D C
mul - ber - ry bush, mul - ber - ry bush.

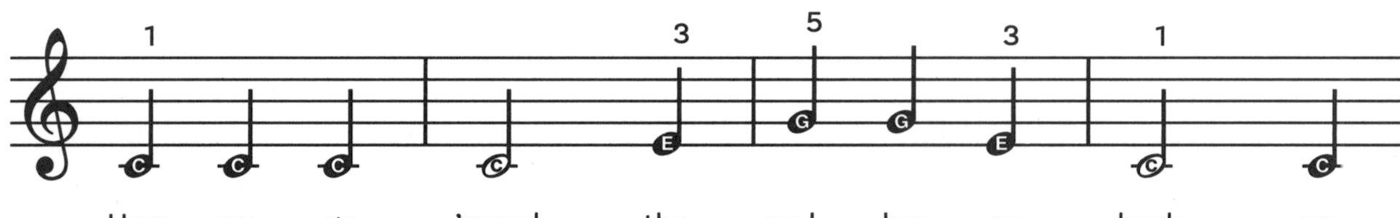

1
3
5
3
1
C C C C E G G E C C
Here we go 'round the mul - ber - ry bush, so

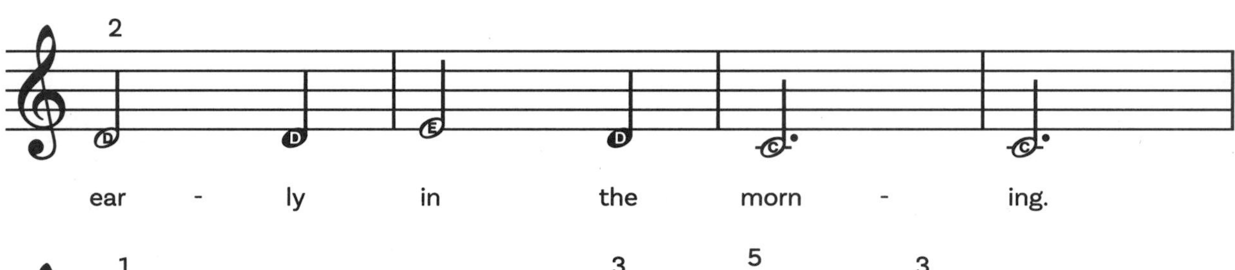

2
D D E D C C
ear - ly in the morn - ing.

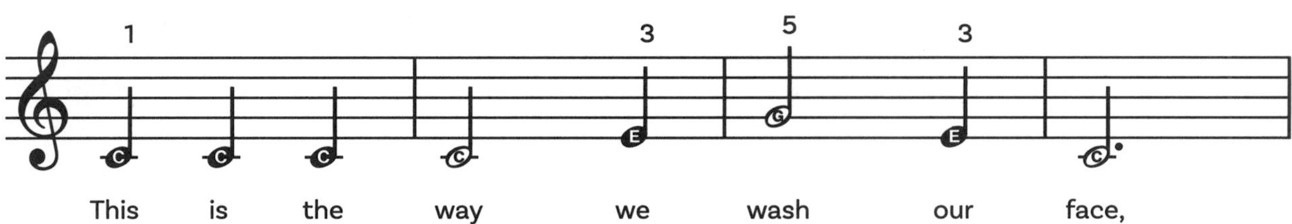

1
3
5
3
C C C C E G E C
This is the way we wash our face,

2
D D D E D C
wash our face, wash our face.

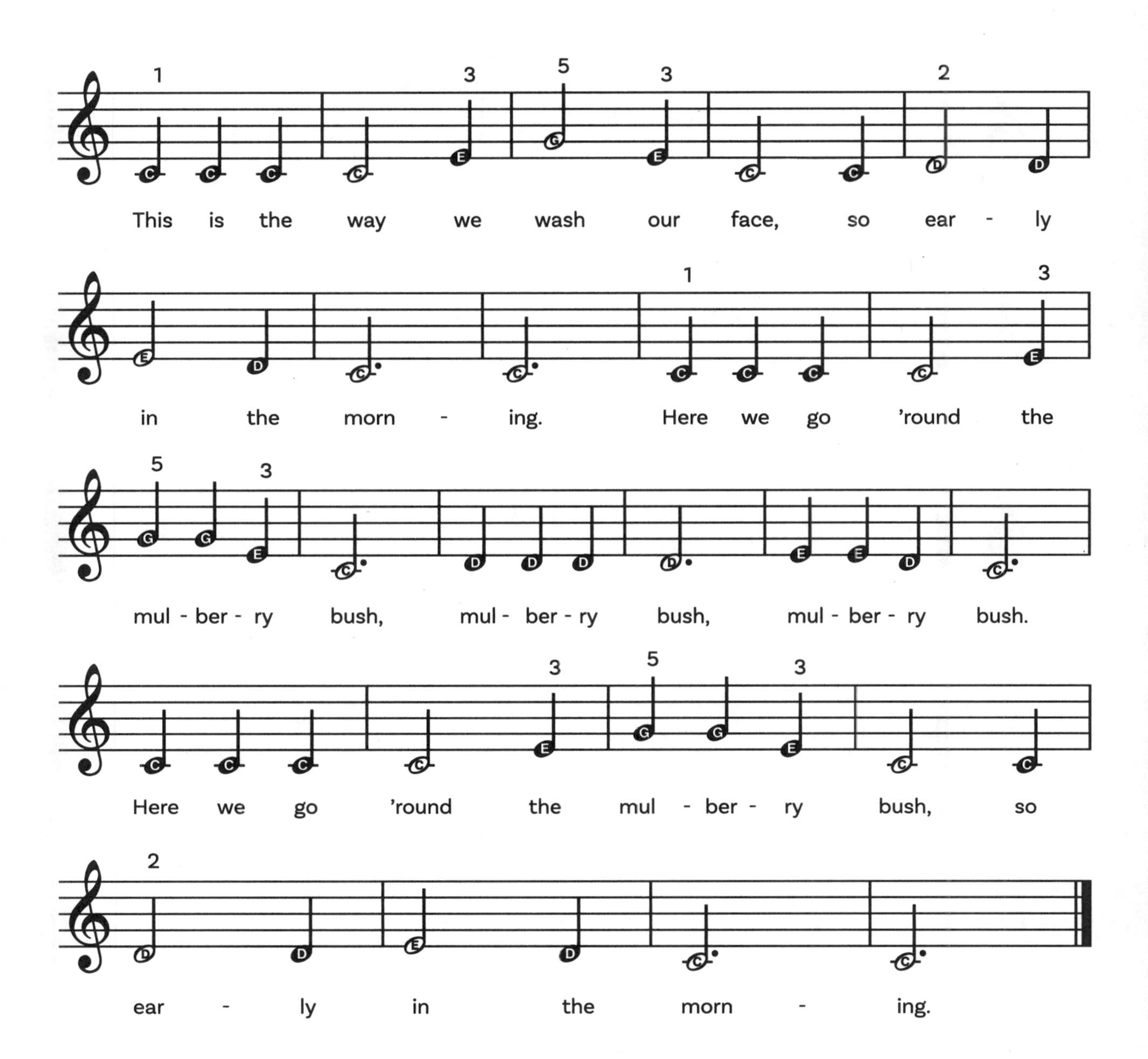
This is the way we wash our face, so ear - ly
in the morn - ing. Here we go 'round the
mul - ber - ry bush, mul - ber - ry bush, mul - ber - ry bush.
Here we go 'round the mul - ber - ry bush, so
ear - ly in the morn - ing.

Shoo, Fly

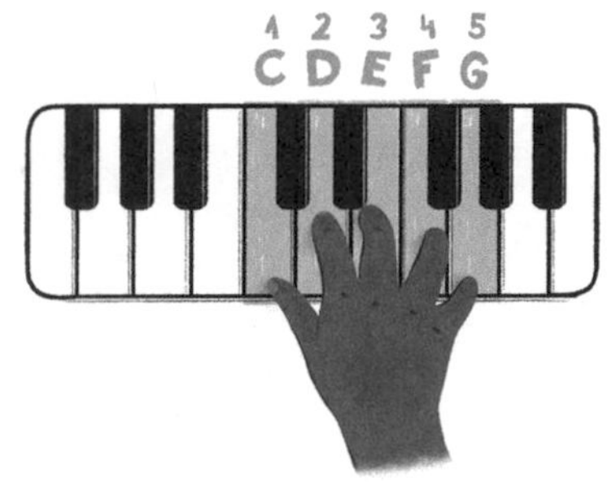
1 2 3 4 5
C D E F G

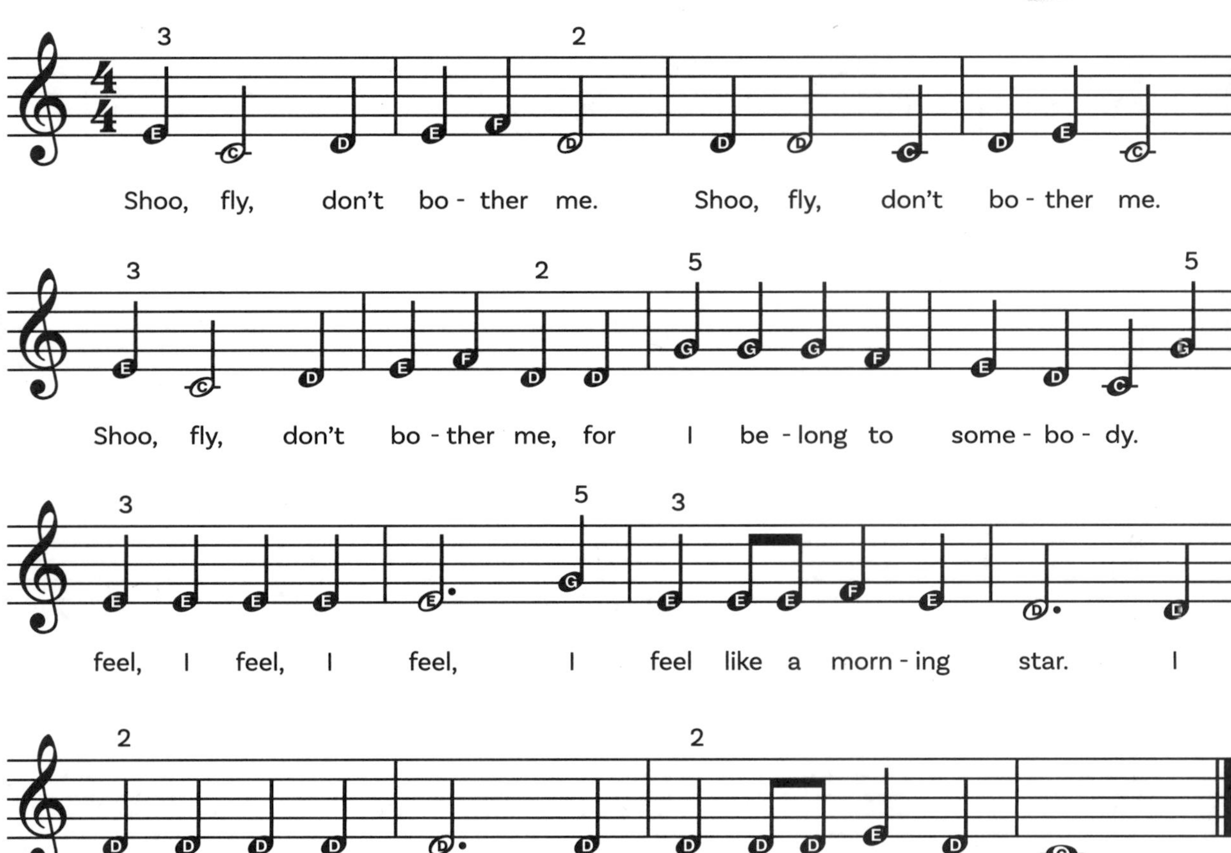
3
2
Shoo, fly, don't bo - ther me. Shoo, fly, don't bo - ther me.
3
2
5
5
Shoo, fly, don't bo - ther me, for I be - long to some - bo - dy.
3
5
3
feel, I feel, I feel, I feel like a morn - ing star. I
2
2
feel, I feel, I feel, I feel like a morn - ing star.

Jingle Bells

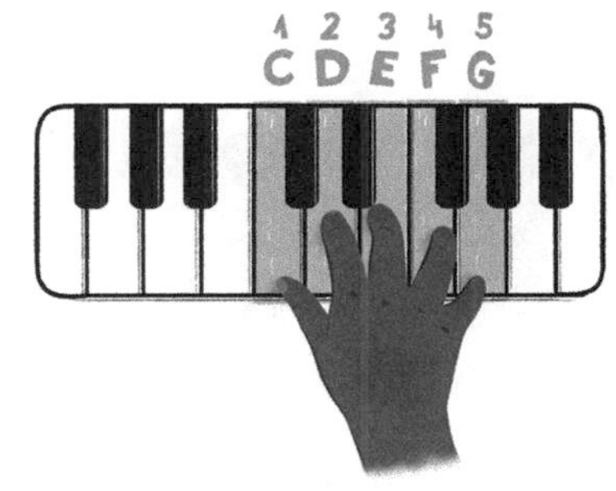
1 2 3 4 5
C D E F G

Jing - le bells, jing - le bells, jing - le all the way.
Oh, what fun it is to ride in a one - horse o - pen sleigh——!
Jing - le bells, jing - le bells, jing - le all the way.
Oh, what fun it is to ride in a one - horse o - pen sleigh!

FUN FACT

The piano is about 5 feet long. That's probably longer than you! It can play both the lowest note and the highest note of any instrument, which means it has the largest range of any instrument.

Skip to My Lou

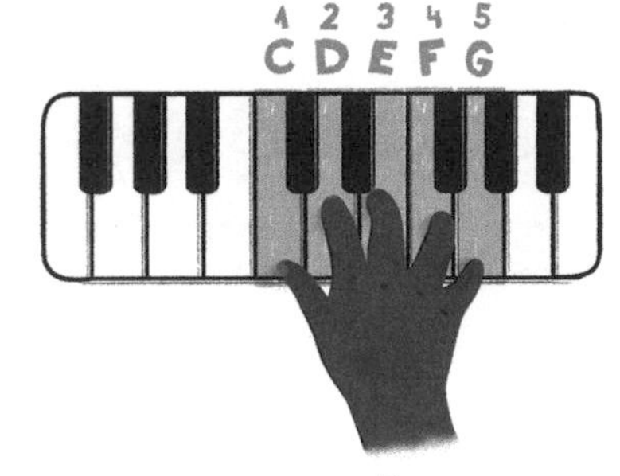
1 2 3 4 5
C D E F G

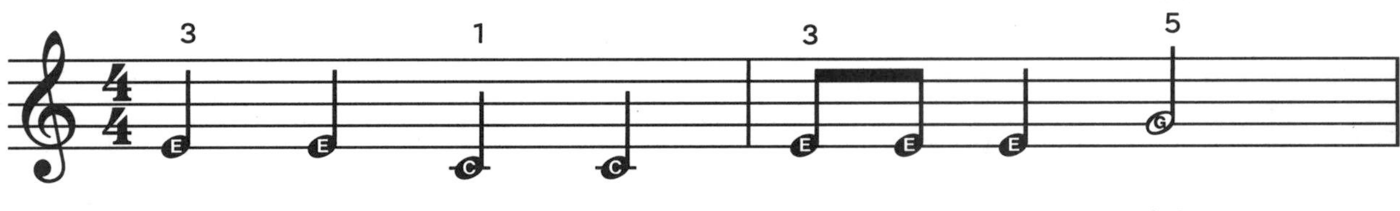

Lost my part - ner, what - 'll I do?

Lost my part - ner, what - 'll I do?

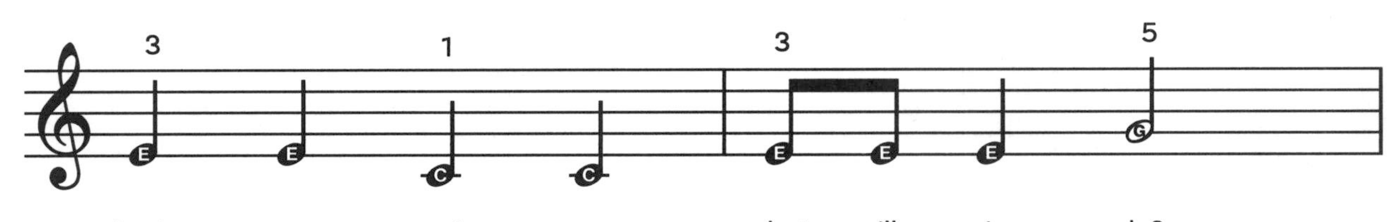

Lost my part - ner, what - 'll I do?

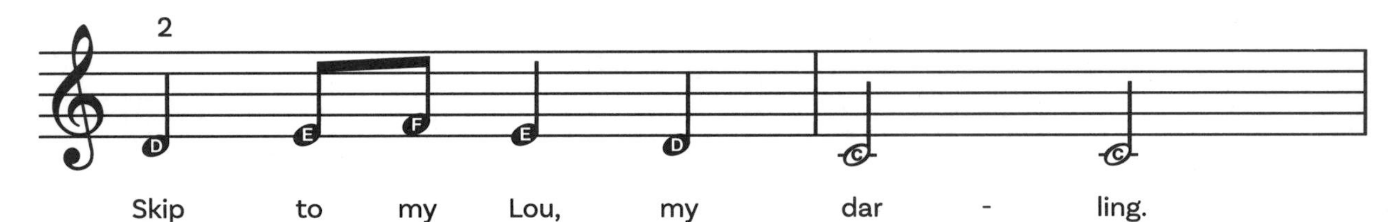

Skip to my Lou, my dar - ling.

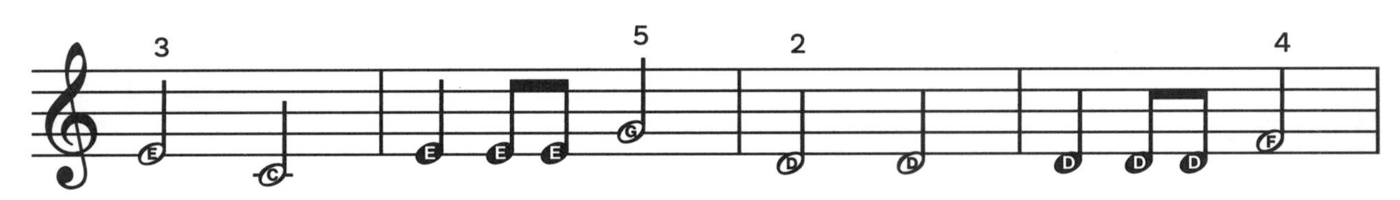

Lou, Lou, skip to my Lou, Lou, Lou, skip to my Lou,

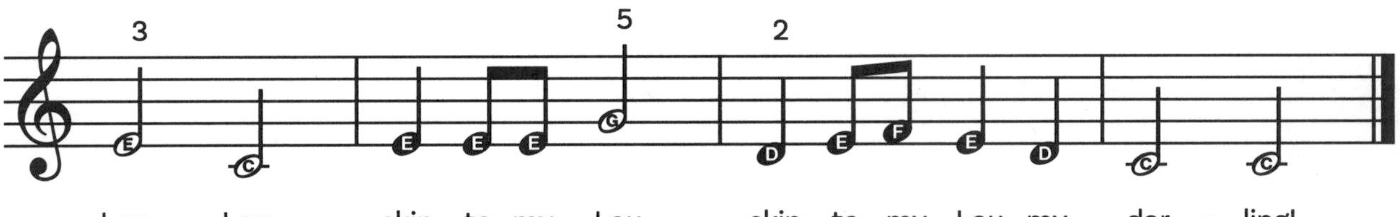

Lou, Lou, skip to my Lou, skip to my Lou, my dar - ling!

To Market, to Market

1 2 3 4 5
C D E F G

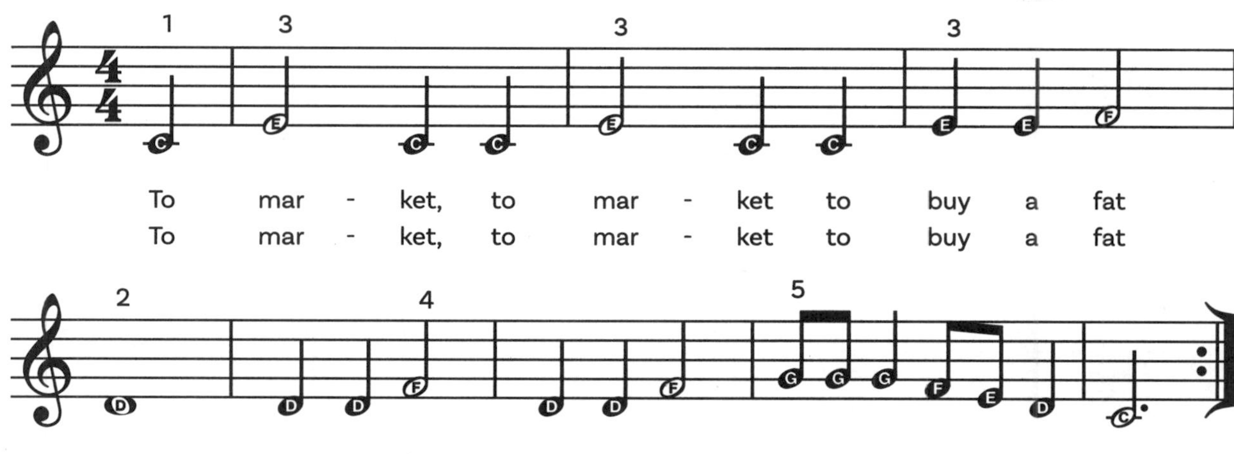
1
3
3
3
To mar - ket, to mar - ket to buy a fat
To mar - ket, to mar - ket to buy a fat
2
4
5
pig. Home a - gain, home a - gain, jig - ge - ty, jig - ge - ty, jig.
hog. Home a - gain, home a - gain, jig - ge - ty, jig - ge - ty, jog.

FUN FACT

Musical instruments are grouped into 5 families—brass, woodwind, string, percussion, and keyboard. Did you know the acoustic piano is considered a percussion instrument, not a keyboard? Percussion instruments are those that make a sound when they are hit. The acoustic piano has hammers that hit the strings to make a sound, so it is a percussion instrument.

Little Liza Jane

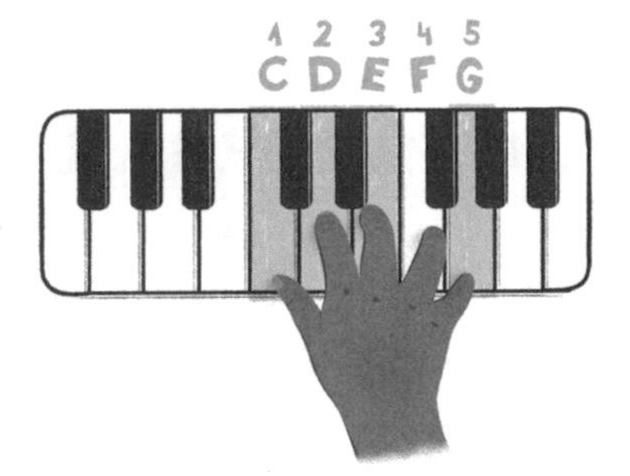
1 2 3 4 5
C D E F G

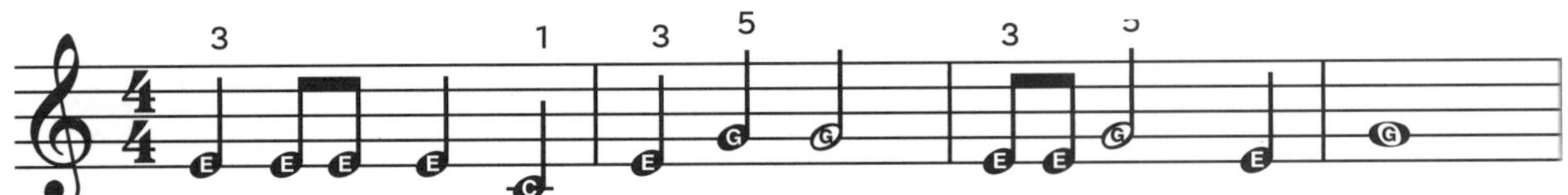
3 1 3 5 3 5
E E E E C E G G E E G E G

I've got a friend in Bal - ti - more, lit - tle Li - za Jane.
I've got a friend in San An - tone, lit - tle Li - za Jane.

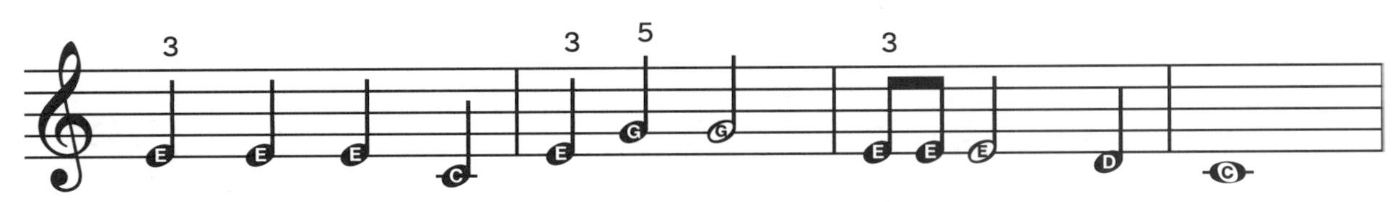
3 3 5 3
E E E C E G G E E E D C

Street - cars run - ning by her door, lit - tle Li - za Jane.
Tum - ble - weeds and cac - tus grow, lit - tle Li - za Jane.

Oh, lit - tle Li - za, lit - tle Li - za Jane,
Oh, lit - tle Li - za, lit - tle Li - za Jane,

5 3
G G G E G E E E D C

oh, lit - tle Li - za, lit - tle Li - za Jane.
oh, lit - tle Li - za, lit - tle Li - za Jane.

Oh My Darling, Clementine

1 2 3 4 5
C D E F G

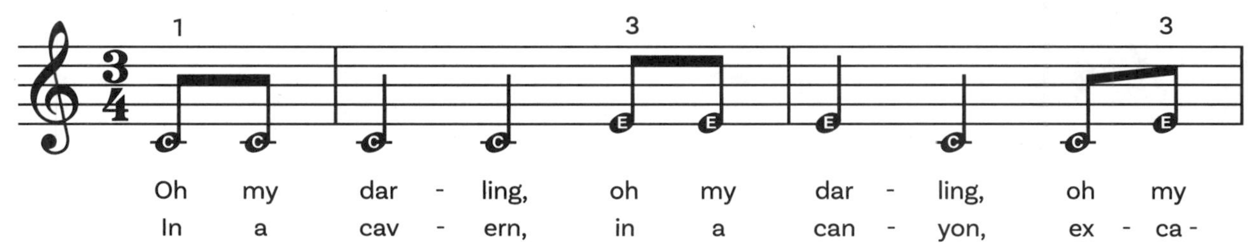

Oh my dar - ling, oh my dar - ling, oh my
In a cav - ern, in a can - yon, ex - ca -

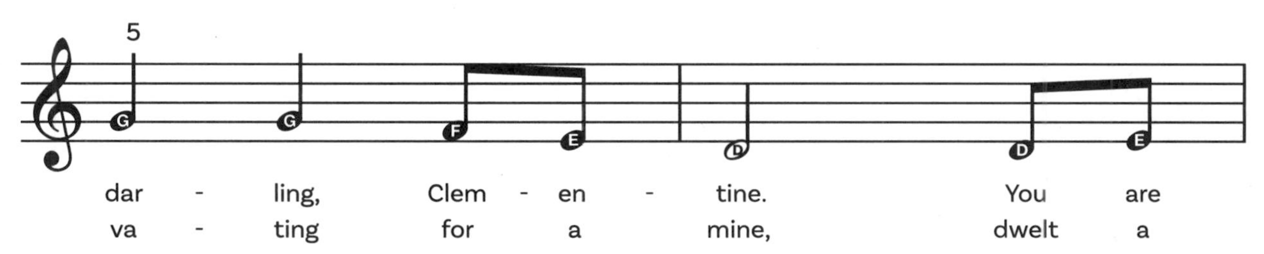

dar - ling, Clem - en - tine. You are
va - ting for a mine, dwelt a

lost and gone for - ev - er, dread - ful sor - ry, Clem - en - tine.
min - er for - ty - ni - ner, and his daugh - ter, Clem - en - tine.

Dry Bones

1 2 3 4 5
C D E F G

1 3
C C C C C C C C E C C
E - ze - ki - el con - nect - ed dem dry bones, E -
2
D D D D D D D E C C
ze - ki - el con - nect - ed dem dry bones, E -
3
C C C C C C C E C C
ze - ki - el con - nect - ed dem dry bones, now
3
E E D D D C C
hear the word of the Lord. Your
3
C C C C C C C E C C
toe bone con - nect - ed to your foot bone, your

2
D D D D D D D E C C
foot bone con - nect - ed to your heel bone, your
C C C C C C C E E C C
3
heel bone con - nect - ed to your ank - le bone, now
3
E E D D D C
hear the word of the Lord.

I'm a Little Teapot

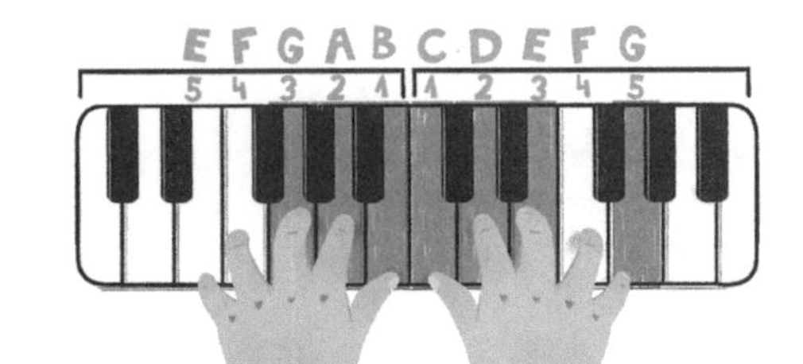

E F G A B C D E F G
5 4 3 2 1 1 2 3 4 5

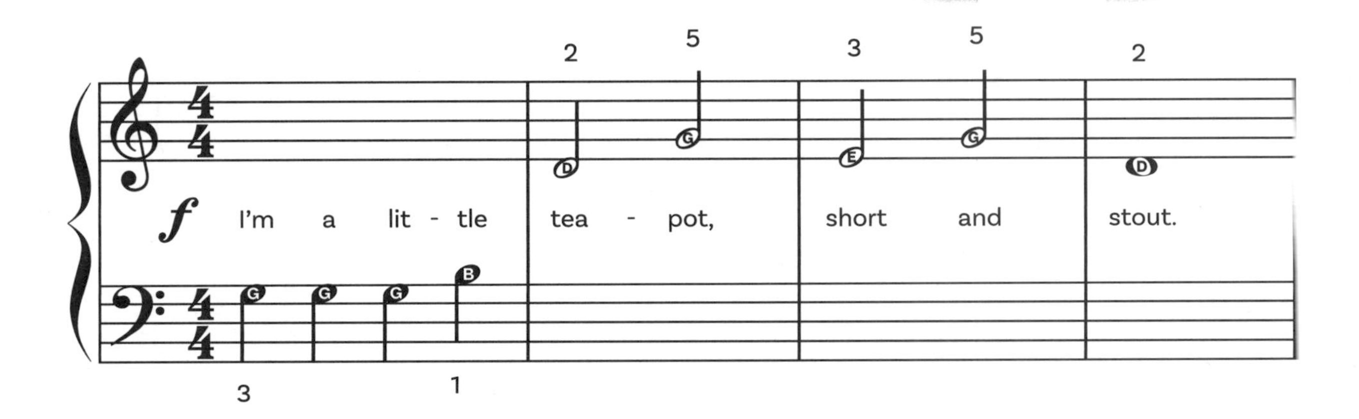

2 5 3 5 2
f I'm a lit - tle tea - pot, short and stout.
G G G B D G E G D
3 1

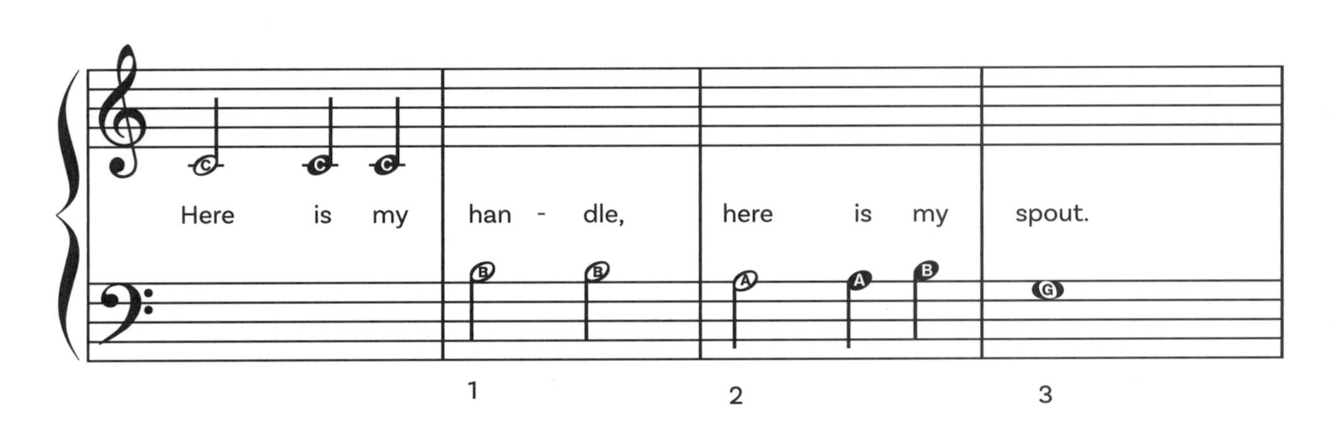

C C C
Here is my han - dle, here is my spout.
B B A A B G
1 2 3

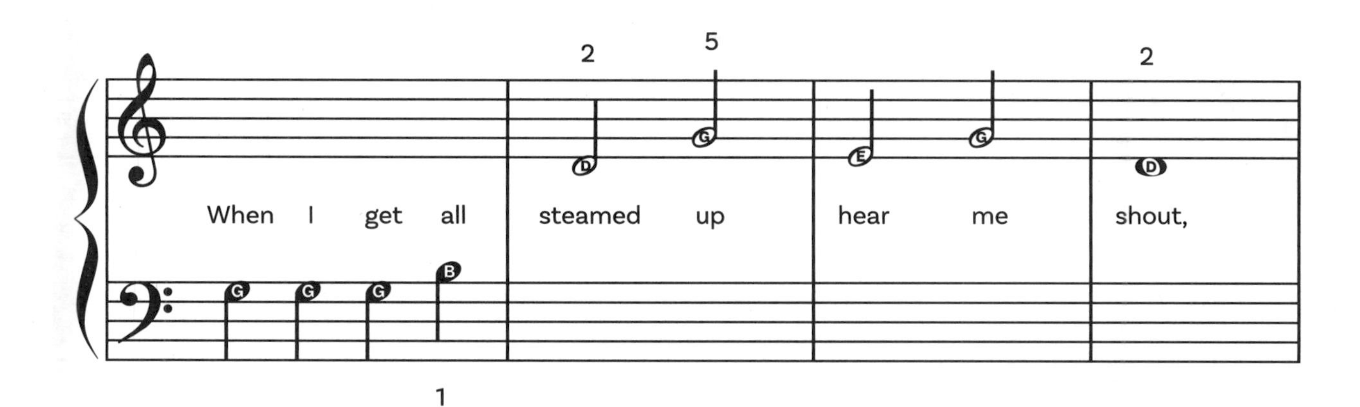
2
5
2
When I get all steamed up hear me shout,
1

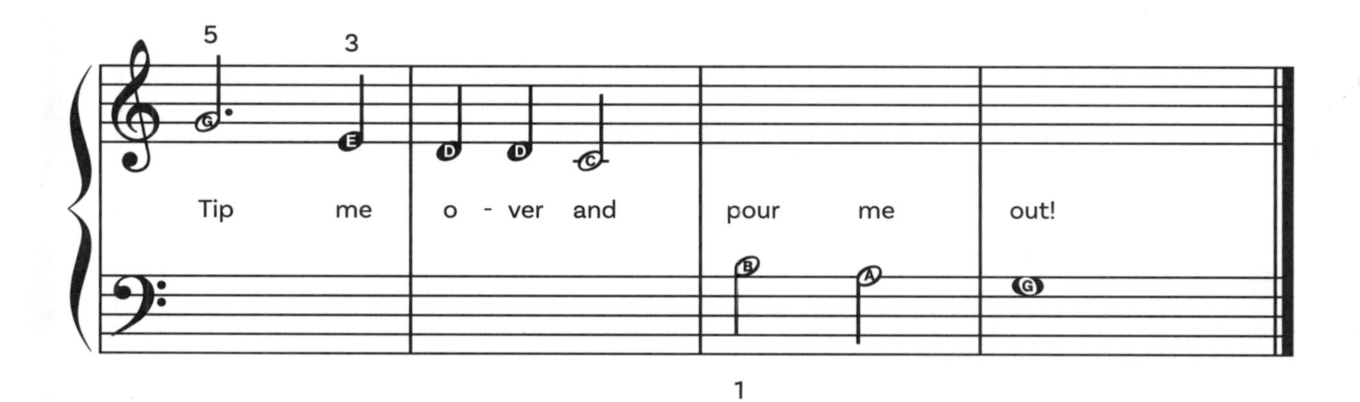
5
3
Tip me o - ver and pour me out!
1

FUN FACT

When you hear a song, you might think hundreds of different notes are being played. But there are only 12 different notes in music! They can sound different based on how low or high they are, what instrument plays them, or what singer sings them.

On Top of Spaghetti

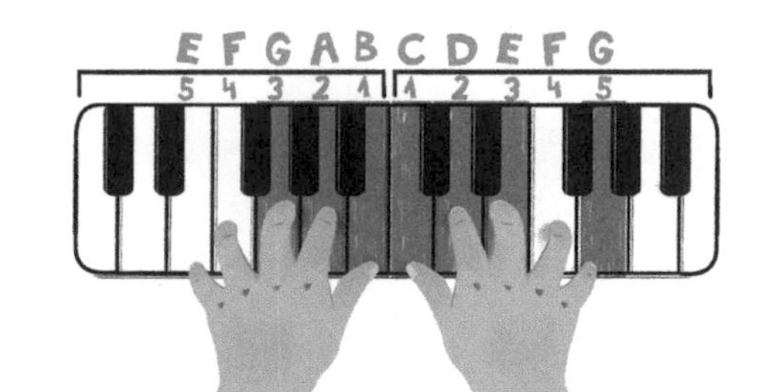
E F G A B C D E F G
5 4 3 2 1 1 2 3 4 5

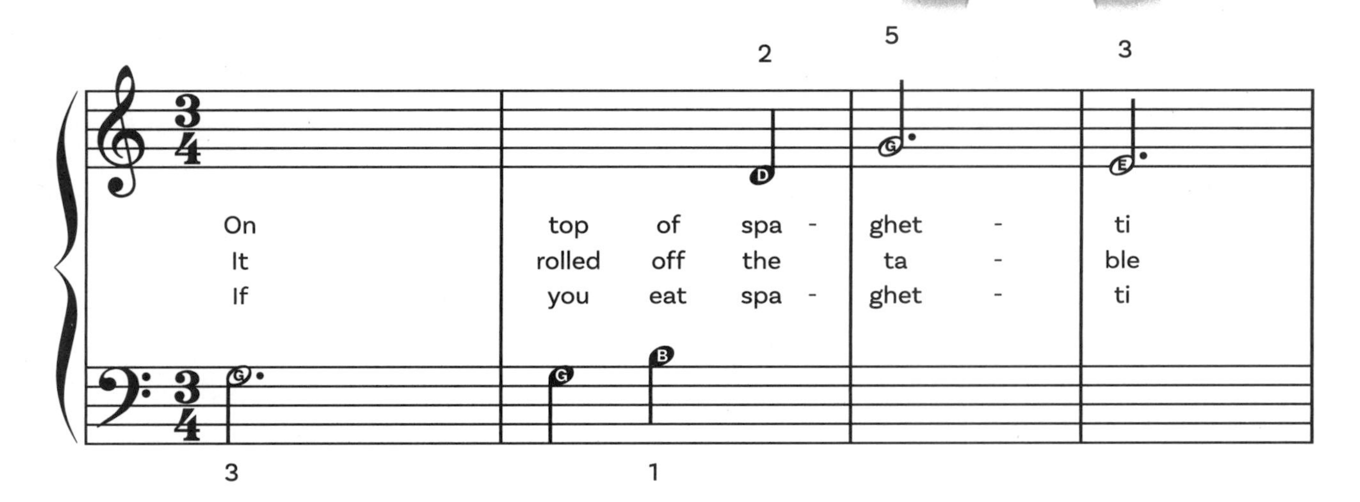
2
5
3
D
G
E
On
It
If
top of spa - ghet - ti
rolled off the ta - ble
you eat spa - ghet - ti
G
G
B
3
1

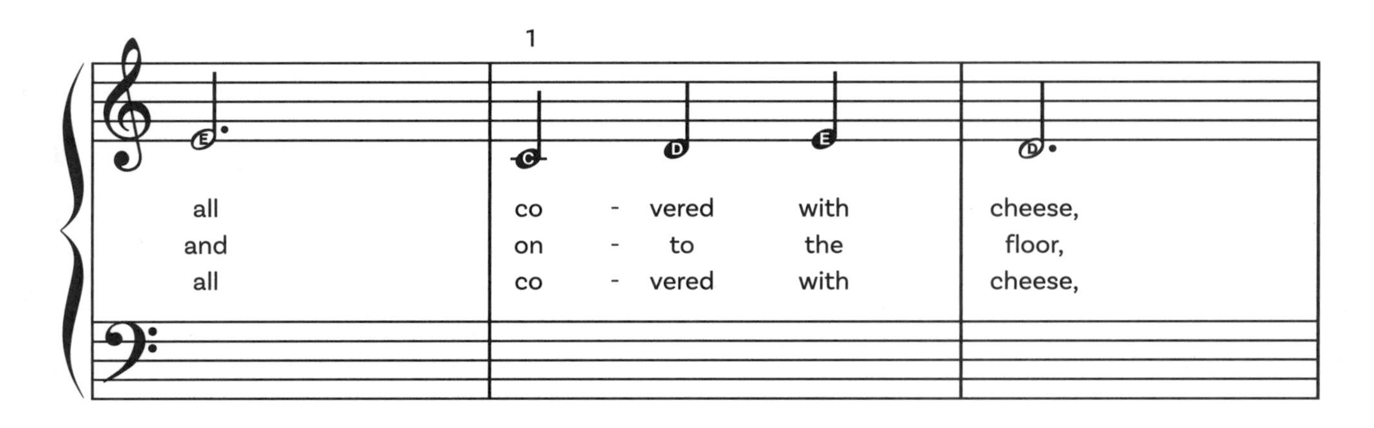
1
E
C
D
E
D
all co - vered with cheese,
and on - to the floor,
all co - vered with cheese,

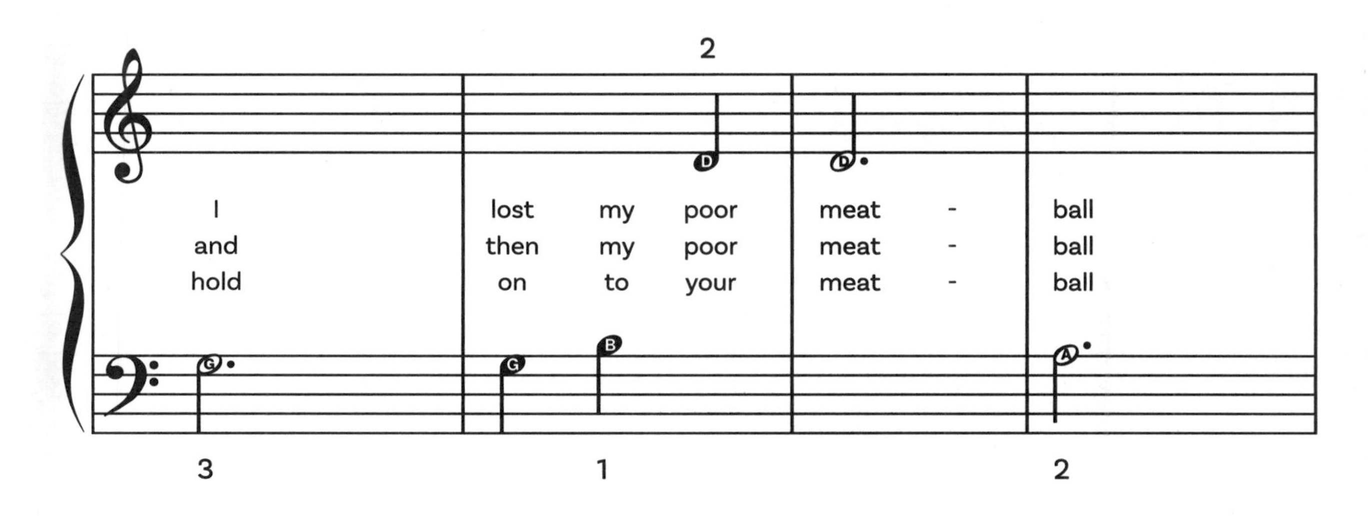
2
I
and
hold
lost my poor meat - ball
then my poor meat - ball
on to your meat - ball
D
D
G
G
B
A
3
1
2

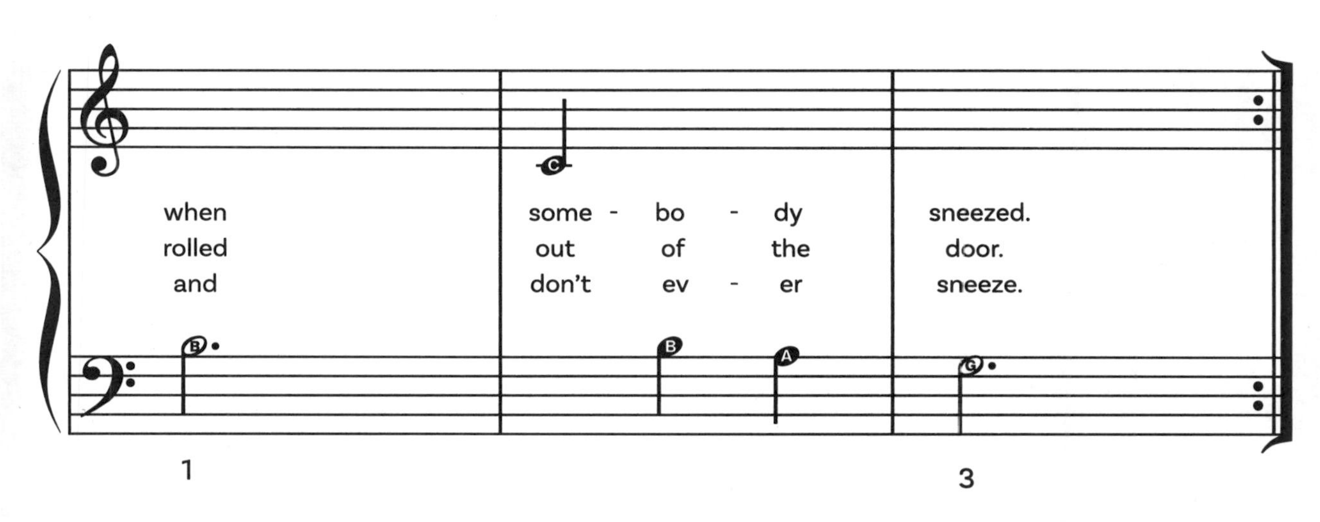
when some - bo - dy sneezed.
rolled out of the door.
and don't ev - er sneeze.
C
B
B
A
G
1
3

The Muffin Man

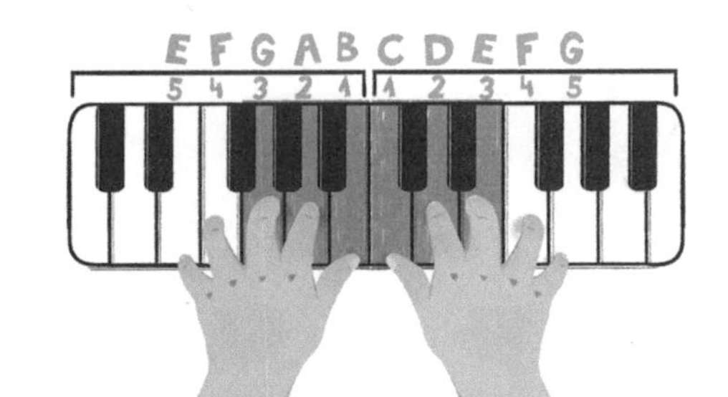
E F G A B C D E F G
5 4 3 2 1 1 2 3 4 5

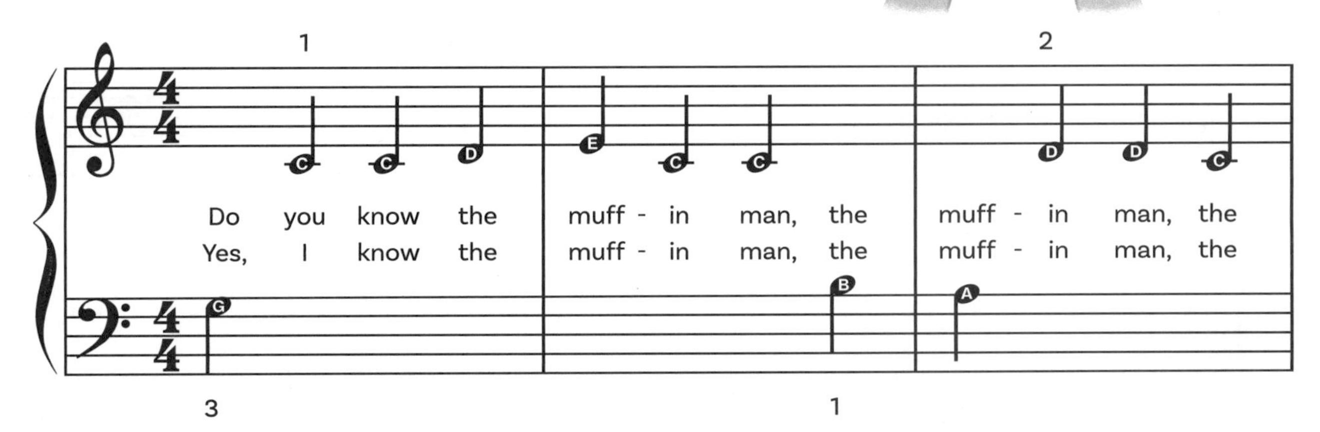
1
2
Do you know the muff - in man, the muff - in man, the
Yes, I know the muff - in man, the muff - in man, the
3
1

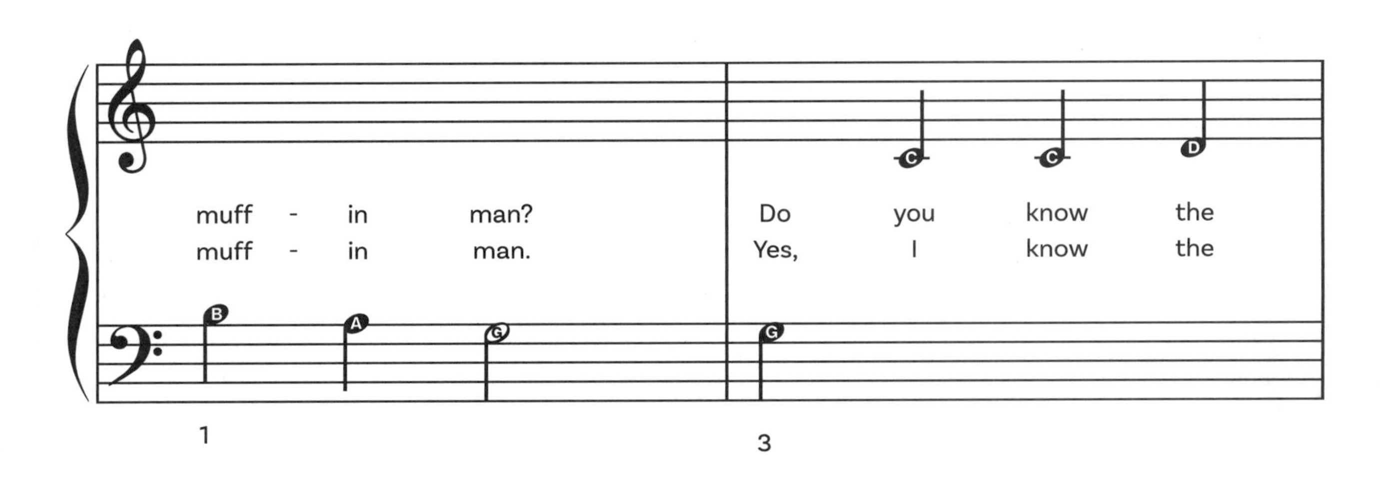
muff - in man? Do you know the
muff - in man. Yes, I know the
1
3

muff - in man who lives on Dru - ry Lane?
muff - in man who lives on Dru - ry Lane.
3

Amazing Grace

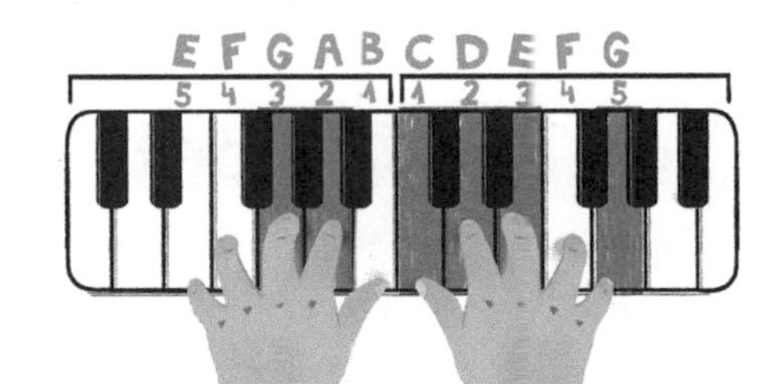

E F G A B C D E F G
5 4 3 2 1 1 2 3 4 5

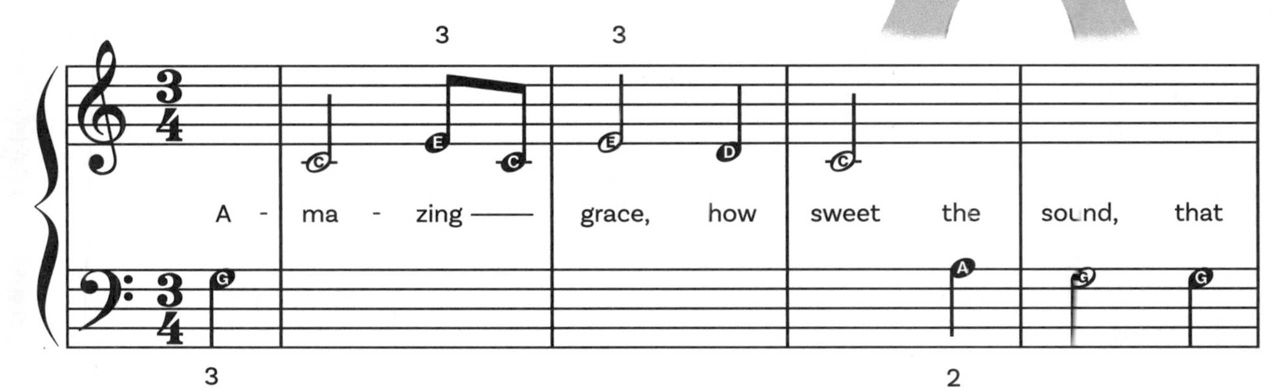

3 3
A - ma - zing grace, how sweet the sound, that

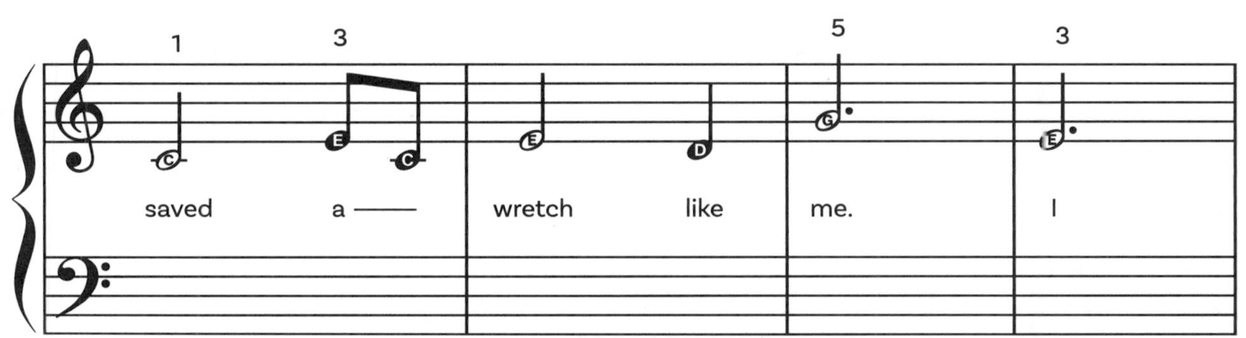

3 1 3 2 5 3
saved a wretch like me. I

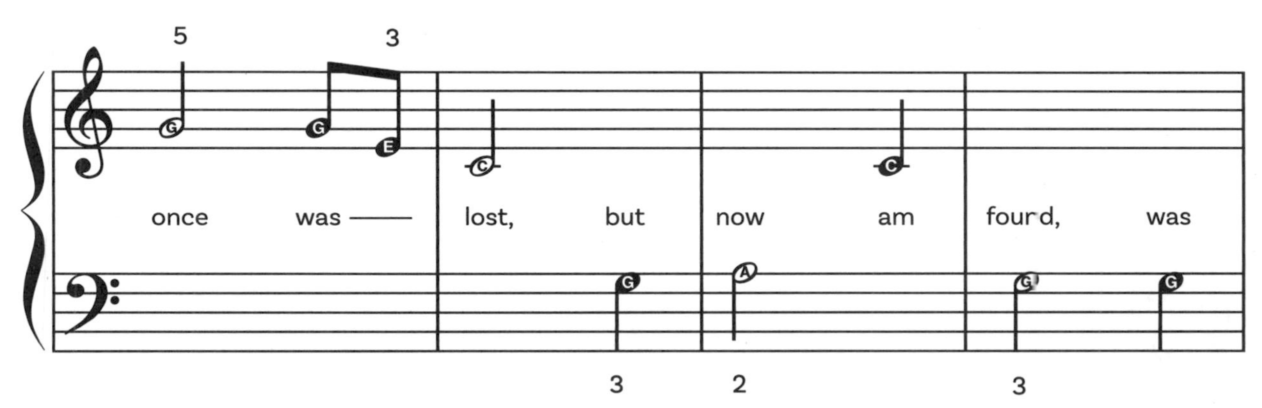

5 3
once was lost, but now am found, was

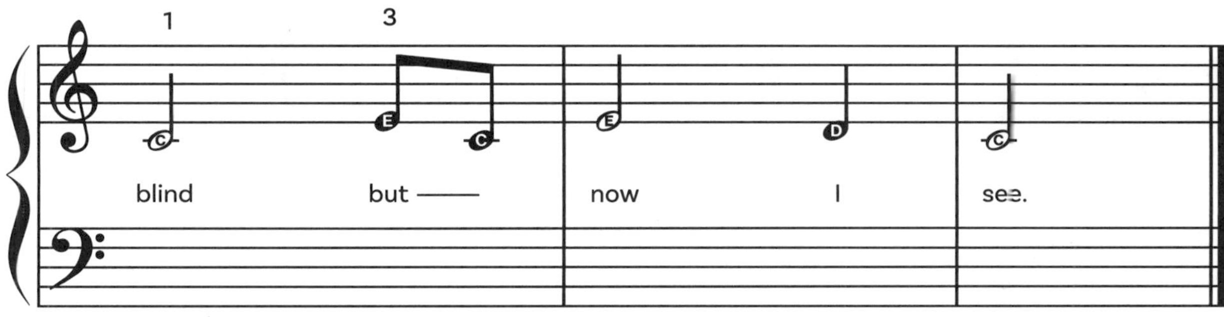

1 3 3 2 3
blind but now I see.

Yankee Doodle

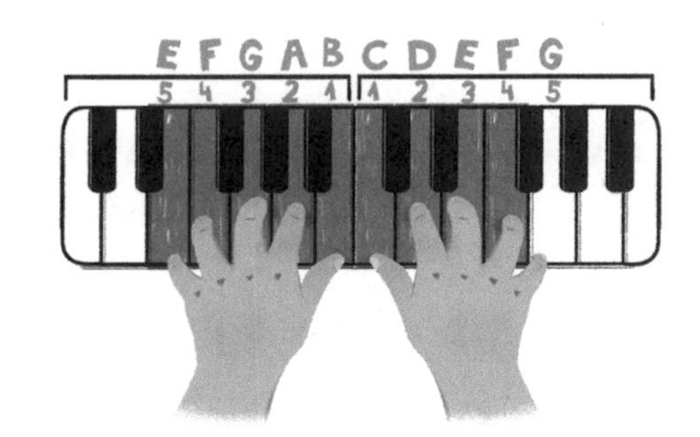
E F G A B C D E F G
5 4 3 2 1 1 2 3 4 5

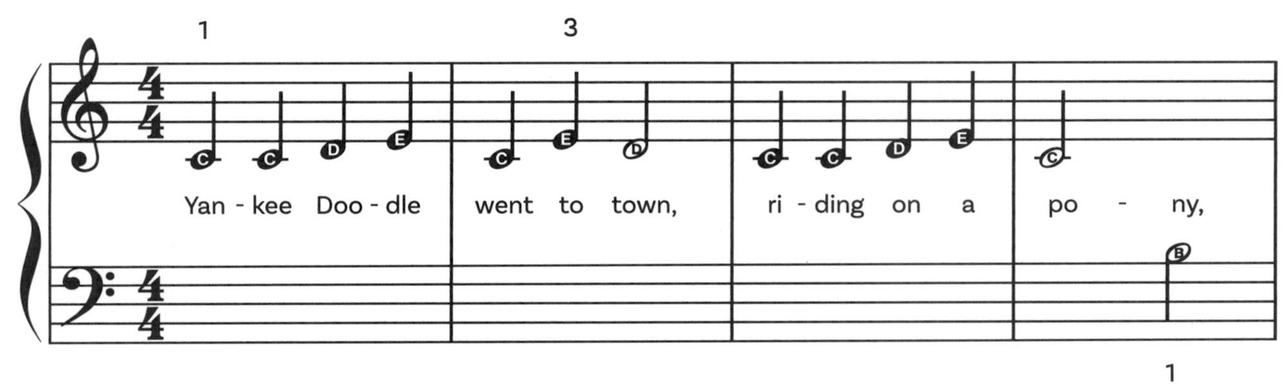
1
3
C C D E
C E D
C C D E
C
Yan - kee Doo - dle
went to town,
ri - ding on a
po - ny,
B
1

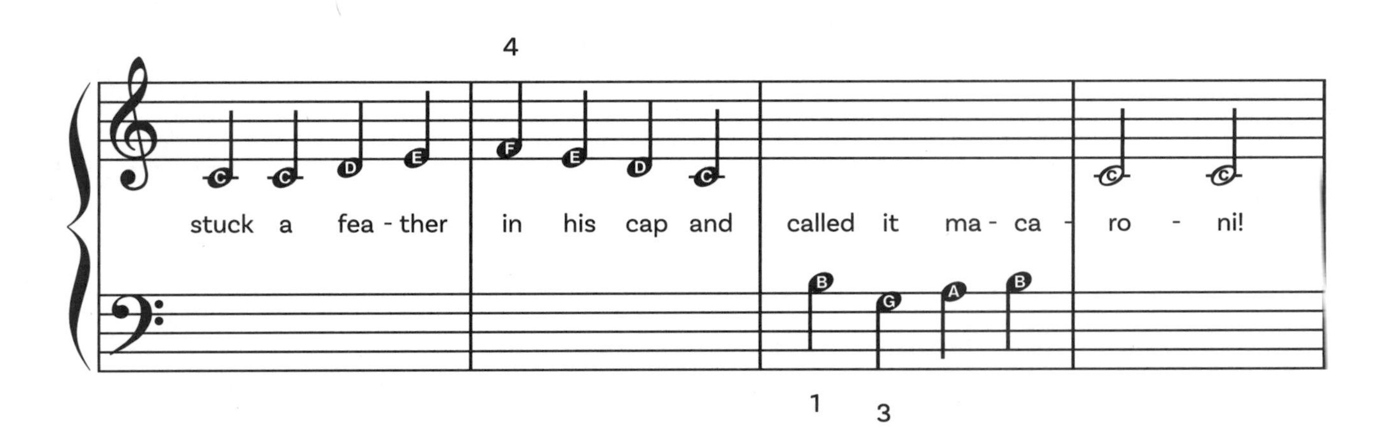
4
C C D E
F E D C
C C
stuck a fea - ther
in his cap and
called it ma - ca -
ro - ni!
B G A B
1 3

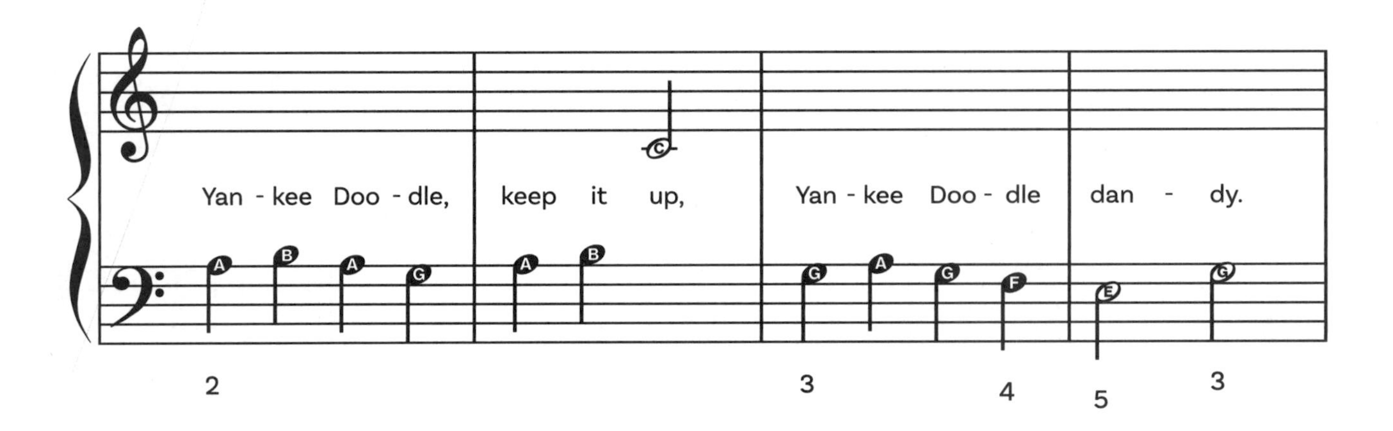
C
Yan - kee Doo - dle,
keep it up,
Yan - kee Doo - dle
dan - dy.
A B A G
A B
G A G F
E G
2
3 4 5 3

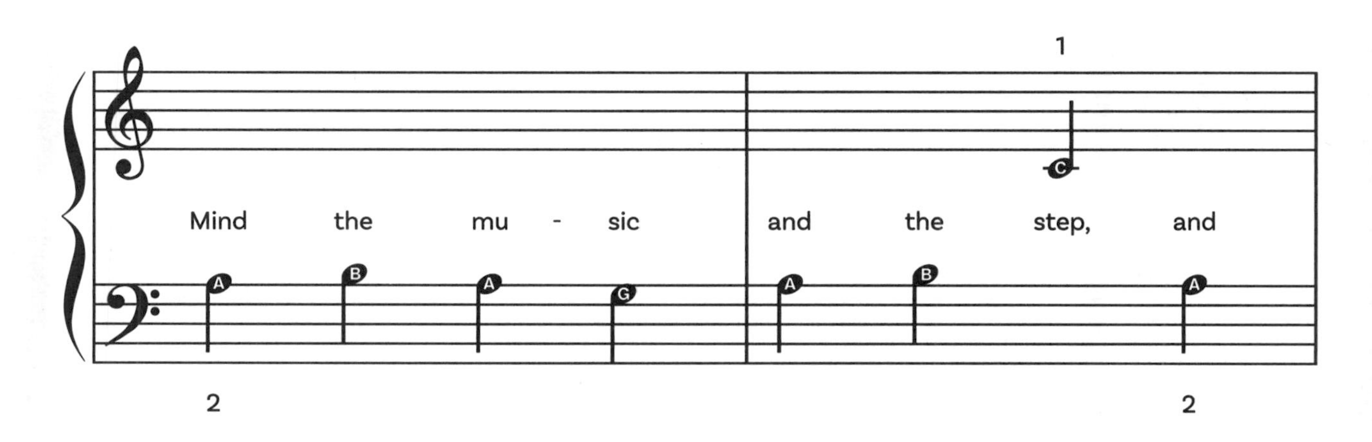
1
Mind the mu - sic and the step, and
A
B
A
G
A
B
C
A
2
2

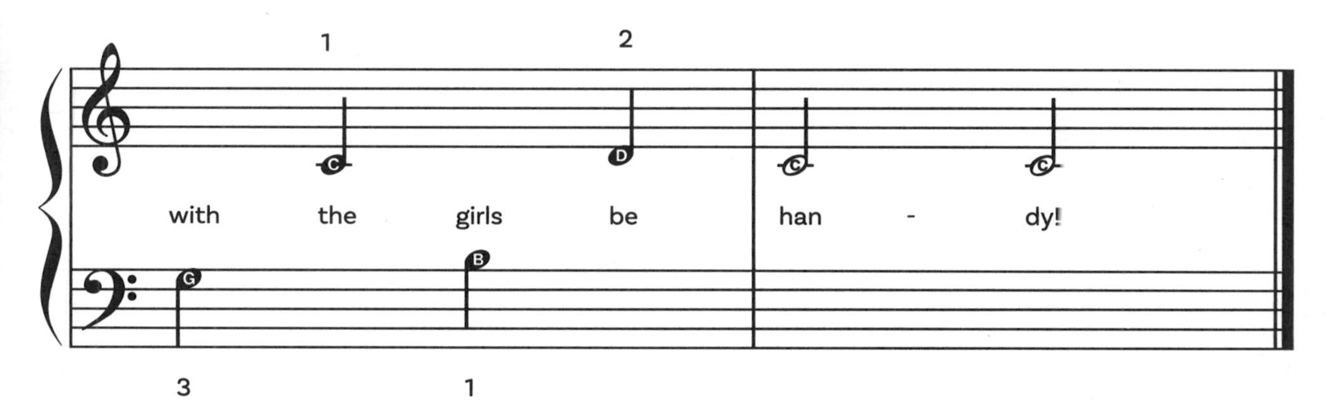
1
2
C
D
C
C
with the girls be han - dy!
G
B
3
1

FUN FACT

The word "pop" in "pop music" is short for "popular." It was first used 70 years ago when rock and roll music was popular with young kids. Now there are hundreds of types of pop music! Hip-hop, dance, K-pop, and country are just a few examples. What kind of pop music do you like?

Over the River and Through the Woods

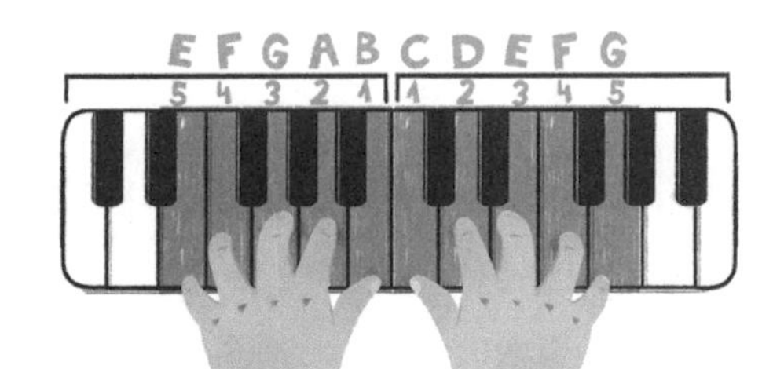
E F G A B C D E F G
5 4 3 2 1 1 2 3 4 5

3
4
O - ver the ri - ver and through the woods to
G G G G E F G G G G
3 5 3
grand - moth - er's house we go. The
C C C B A G G
1 3
3
horse knows the way to car - ry the sleigh through
E
F F F F F E E E E
4 5 5
5
white and drift - ed snow.
D D D E D G

O - ver the ri - ver and through the
3
5
woods, oh, how the wind does blow!
1
3
It stings the toes and bites the
3
1
1
3
5
nose, as o - ver the ground we go.

If You're Happy and You Know It

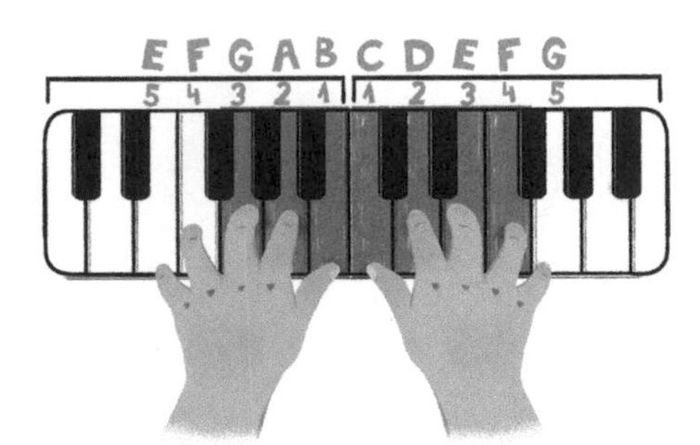

E F G A B C D E F G
5 4 3 2 1 1 2 3 4 5

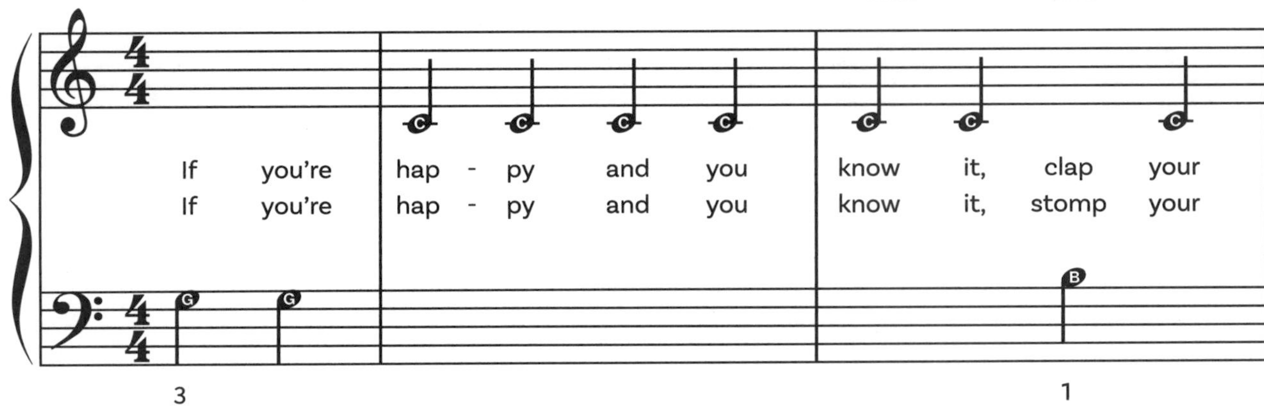

If you're hap - py and you know it, clap your
If you're hap - py and you know it, stomp your
C C C C C C C
G G B
3
1

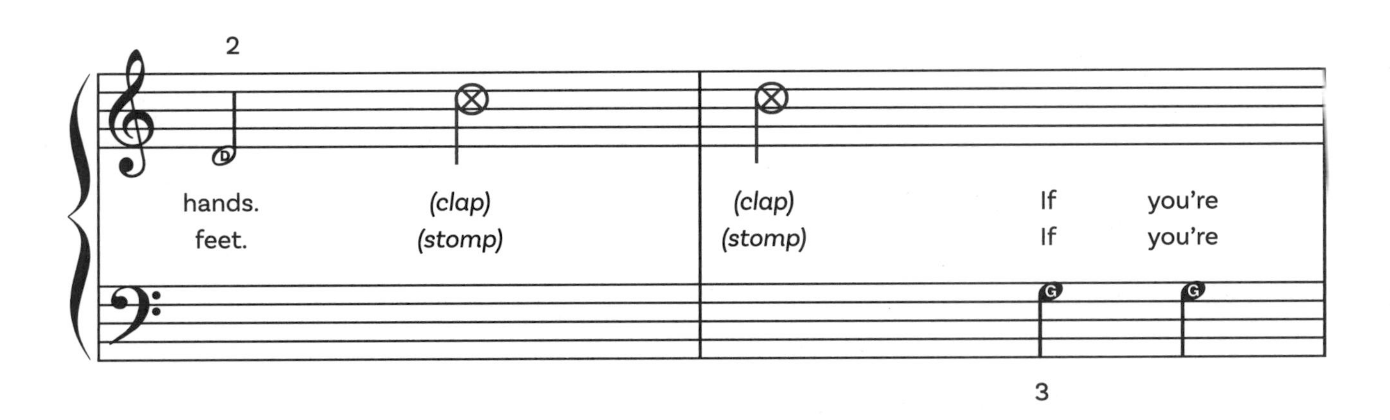

2
hands. (clap) (clap) If you're
feet. (stomp) (stomp) If you're
D
G G
3

2
3
hap - py and you know it, clap your hands. (clap) (clap). If you're
hap - py and you know it, stomp your feet. (stomp) (stomp). If you're
D D D D D D C D E E E

4 3 3
hap - py and you know it, and you real - ly want to show it, if you're
hap - py and you know it, and you real - ly want to show it, if you're
2

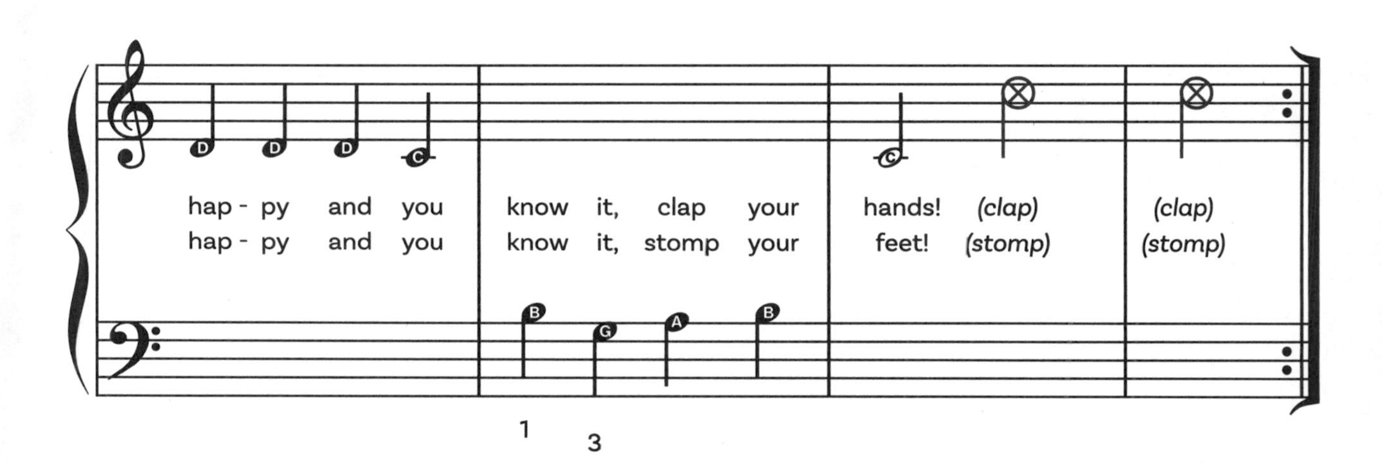
hap - py and you know it, clap your hands! (clap) (clap)
hap - py and you know it, stomp your feet! (stomp) (stomp)
1 3

FUN FACT

The members of the popular British rock band the Beatles could NOT read any music. The songs just came to them in their heads, and they played.

Bingo

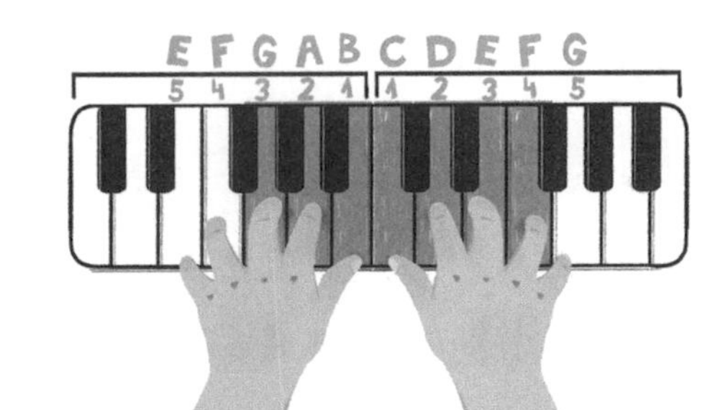
E F G A B C D E F G
5 4 3 2 1 1 2 3 4 5

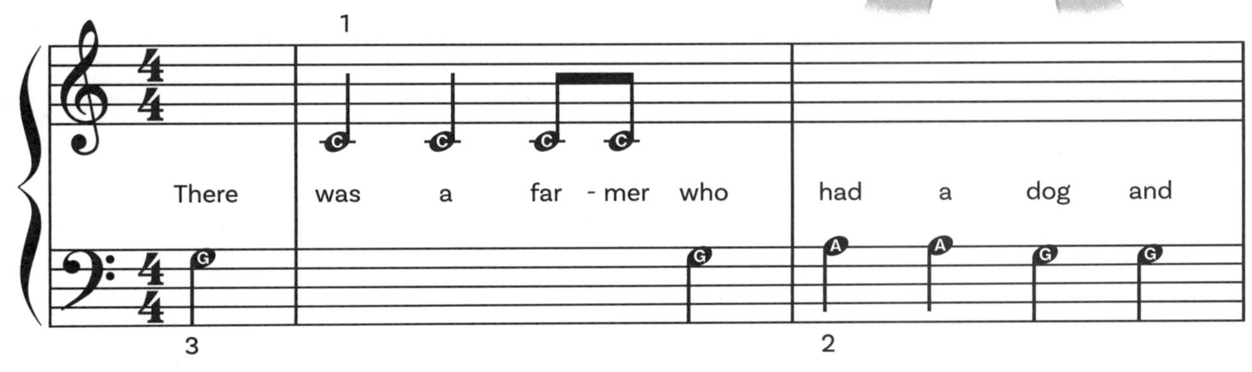
1
C C C C
There was a far - mer who had a dog and
G G A A G G
3
2

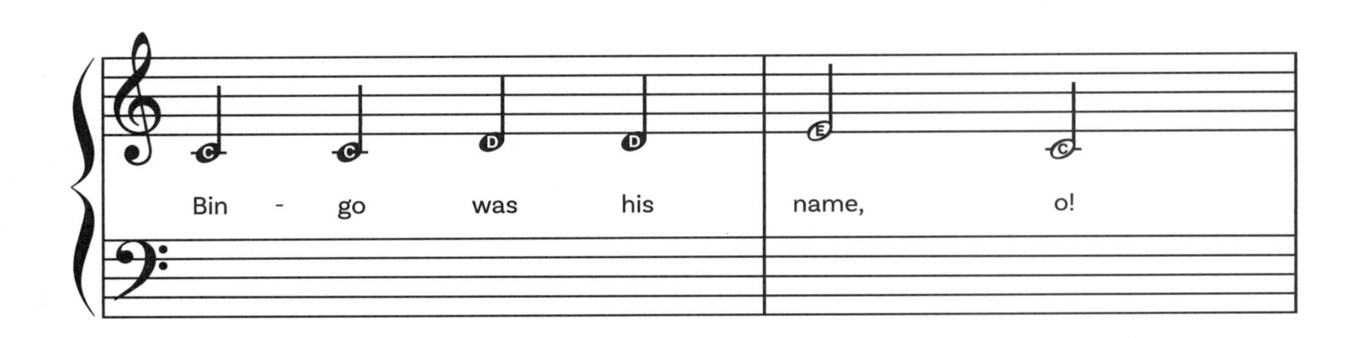
C C D D E C
Bin - go was his name, o!

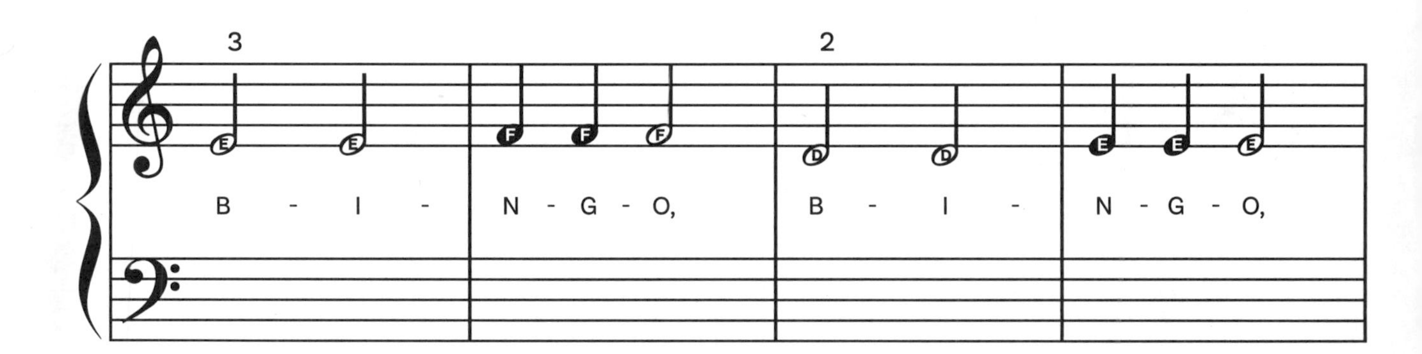
3
2
E E F F F D D E E E
B - I - N - G - O, B - I - N - G - O,

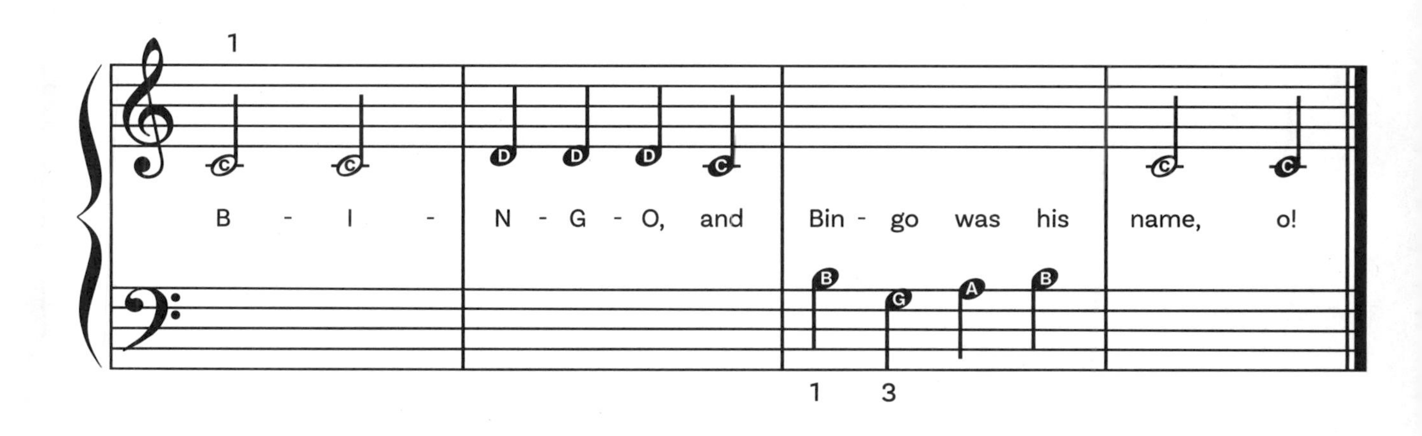
1
C C D D D C
B - I - N - G - O, and Bin - go was his name, o! C C
B G A B
1 3

FUN FACT

In the 1700s, a composer wrote a song called "Cat's Fugue" by writing down the notes his cat played as it walked across the piano keys. So even animals can play music!

Old MacDonald

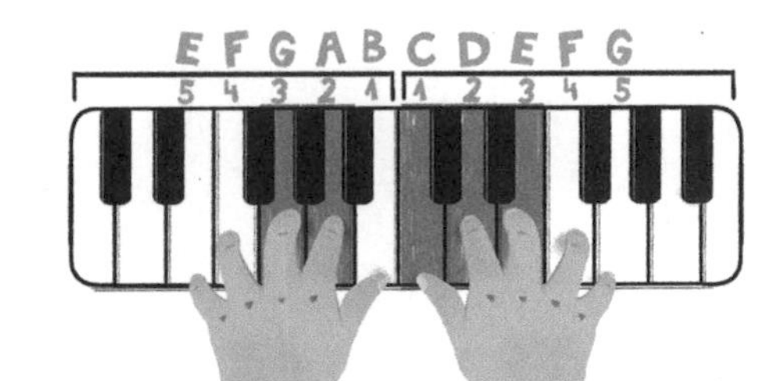
E F G A B C D E F G
5 4 3 2 1 1 2 3 4 5

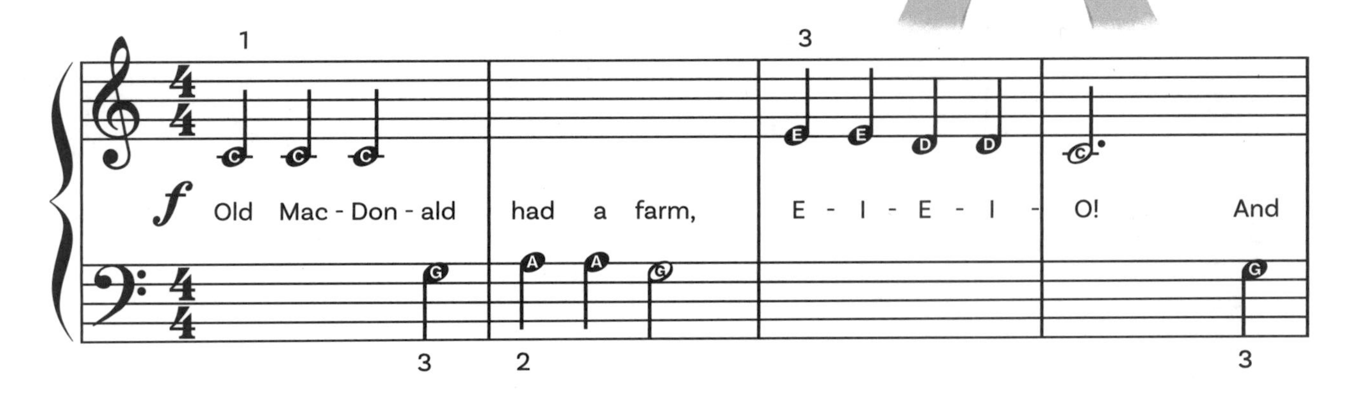
1
3
C C C
E E D D
C
f Old Mac - Don - ald
had a farm,
E - I - E - I -
O! And
G
A A G
G
3 2
3

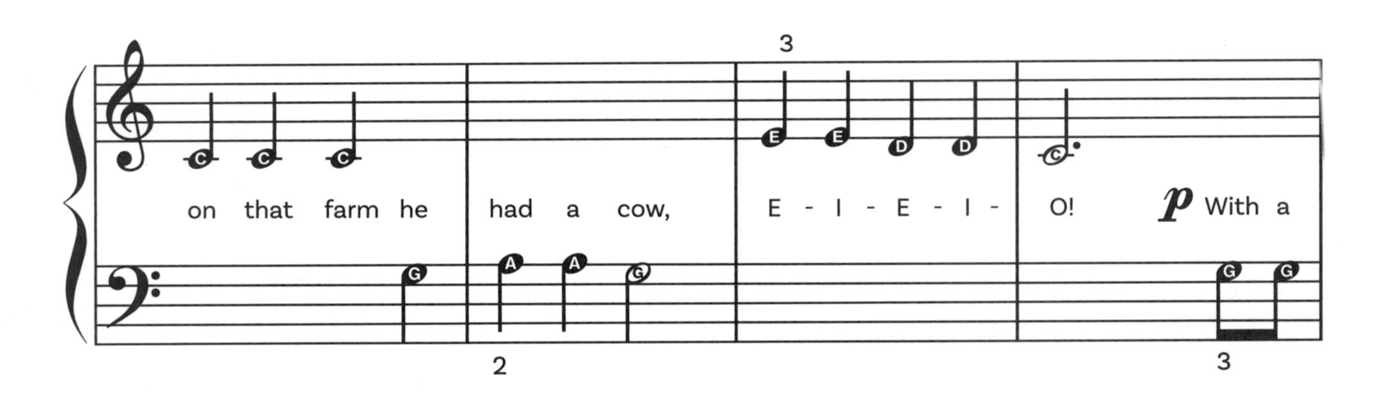
3
C C C
E E D D
C
on that farm he
had a cow,
E - I - E - I -
O! p With a
G
A A G
G G
2
3

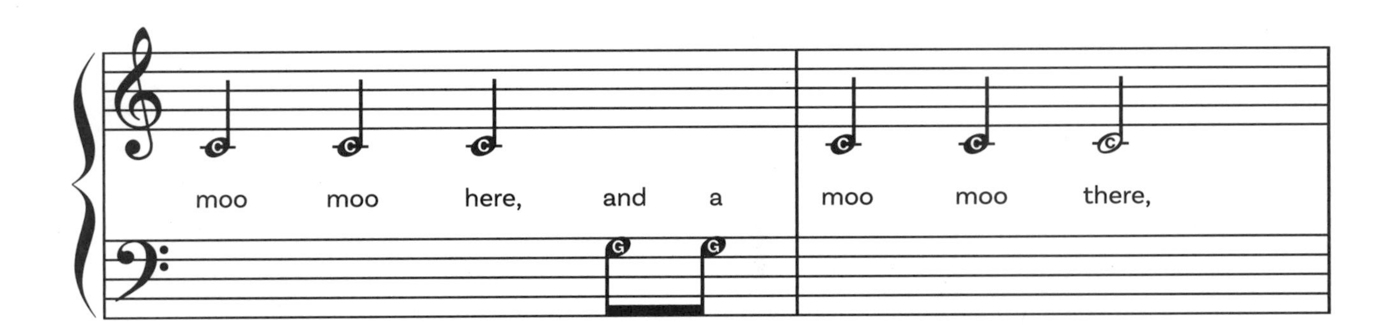
C C C
C C C
moo moo here, and a
moo moo there,
G G

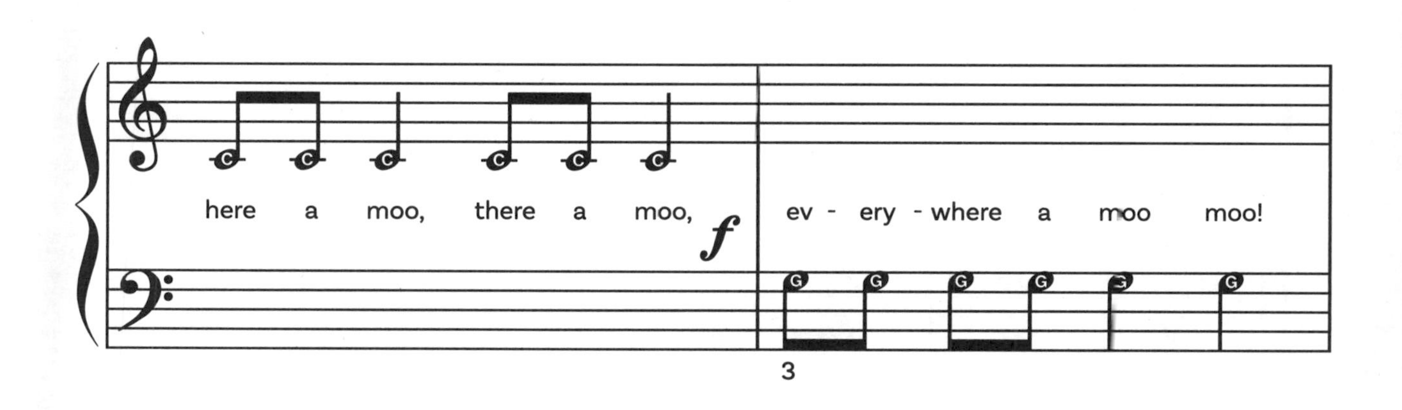
C C C C C C
here a moo, there a moo,
f
ev - ery - where a moo moo!
G G G G G G
3

C C C
Old Mac - Don - ald
had a farm,
E E D D
E - I - E - I -
C
O!
G
A A G
2

Happy Birthday

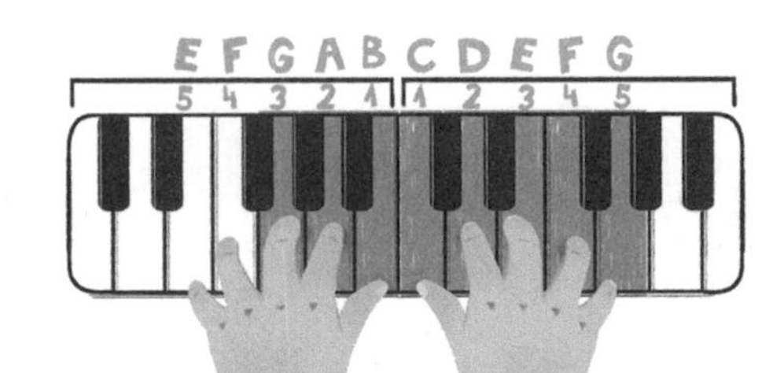
E F G A B C D E F G
5 4 3 2 1 1 2 3 4 5

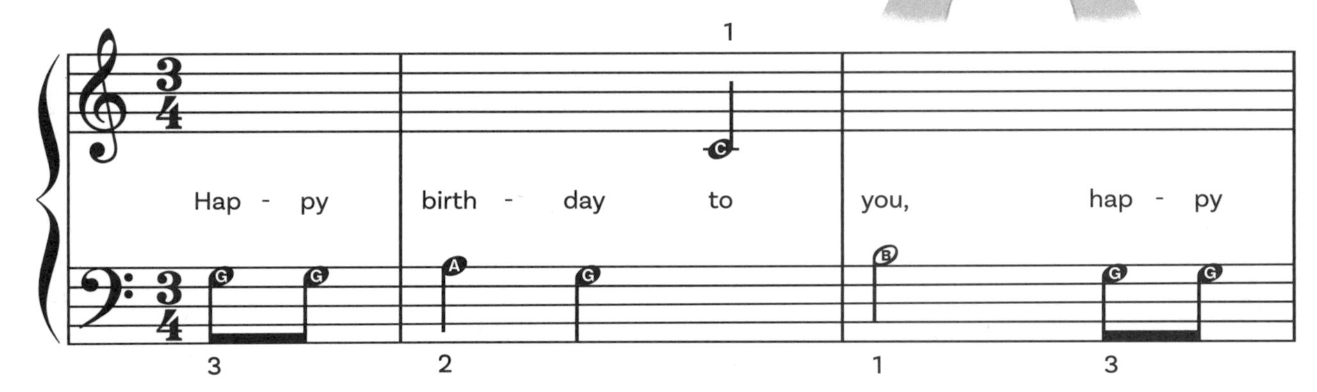
1
Hap - py
birth - day to
you, hap - py
3
2
1
3

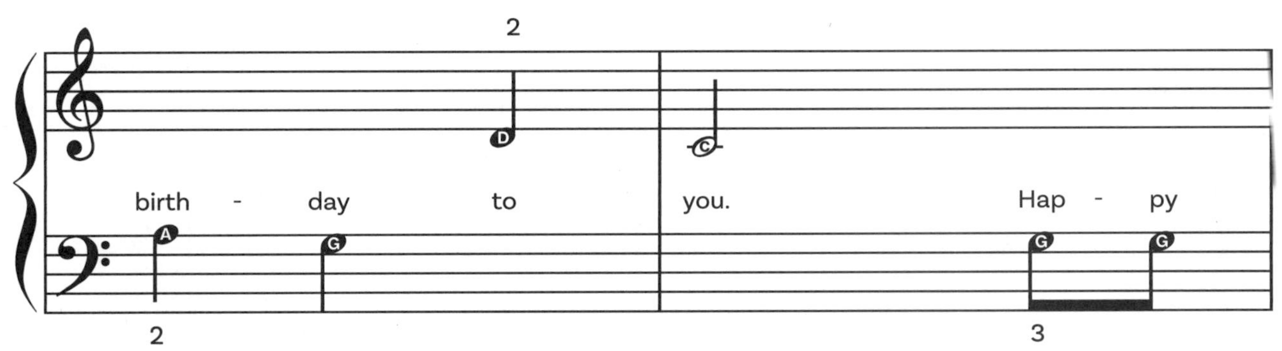
2
birth - day to
you. Hap - py
2
3

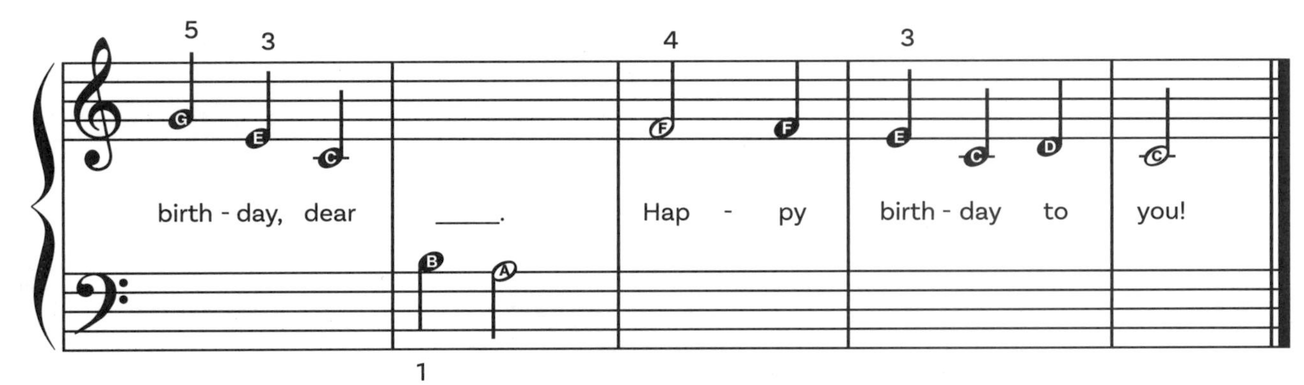
5
3
4
3
birth - day, dear
_____.
Hap - py
birth - day to
you!
1

Twinkle, Twinkle, Little Star

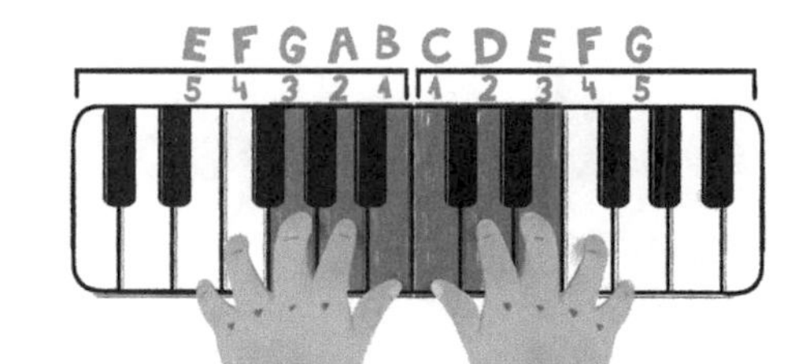
E F G A B C D E F G
5 4 3 2 1 1 2 3 4 5

2
D D E E D C C
G G B B A A G
3 1 2
2 2
D D C C D D C C
B B A B B A
1 1
2
D D E E D C C
G G B B A A G
3 1 2 3

FUN FACT

The famous Austrian composer Wolfgang Amadeus Mozart wrote several variations of this melody in the 18th century. The lyrics that we know and love came later—from a poem by Jane Taylor.

A Little Night Music

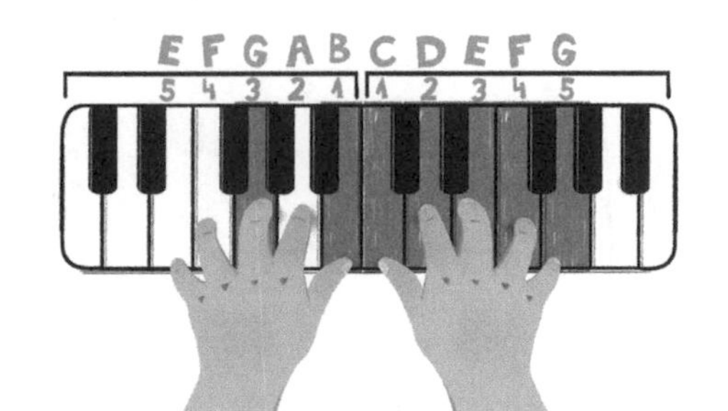
E F G A B C D E F G
5 4 3 2 1 1 2 3 4 5

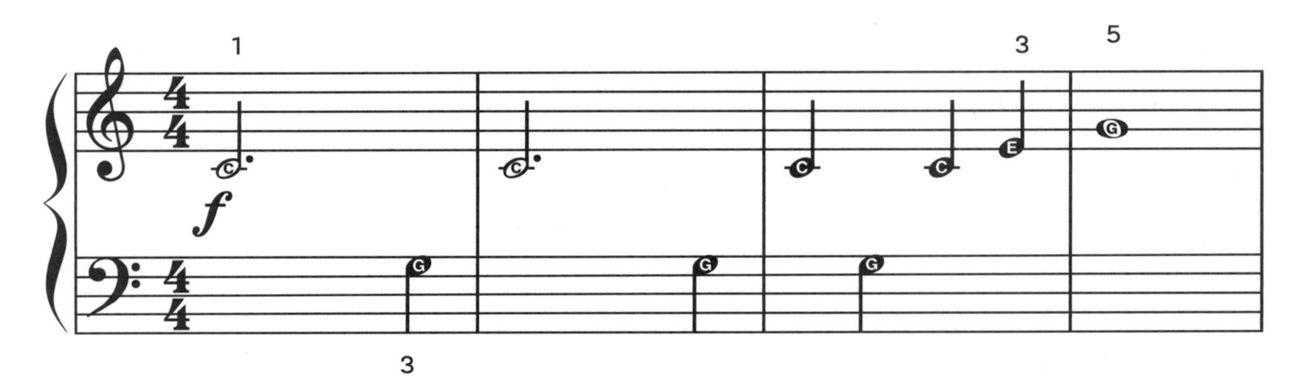
1
3
5
f
C C C C E G
G G G
3

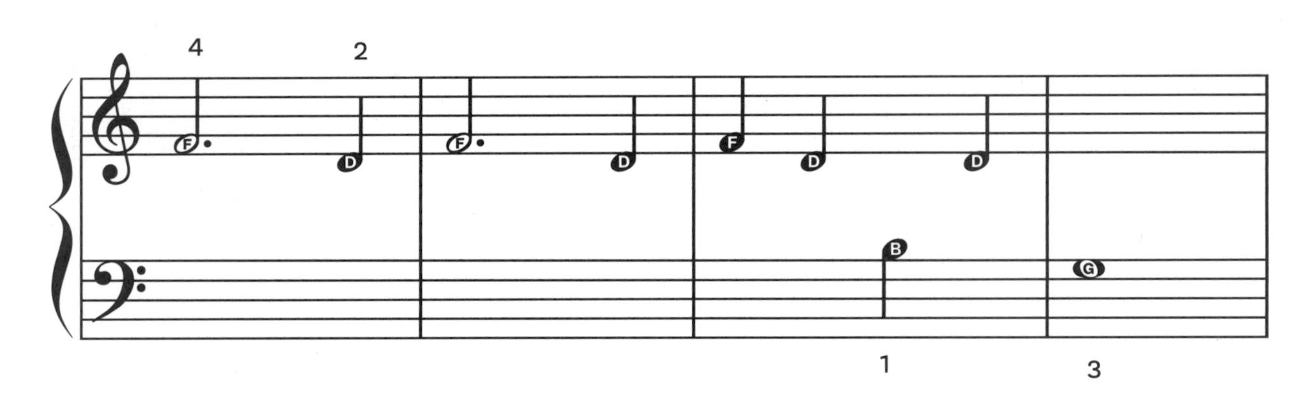
4
2
F D F D F D D
B G
1
3

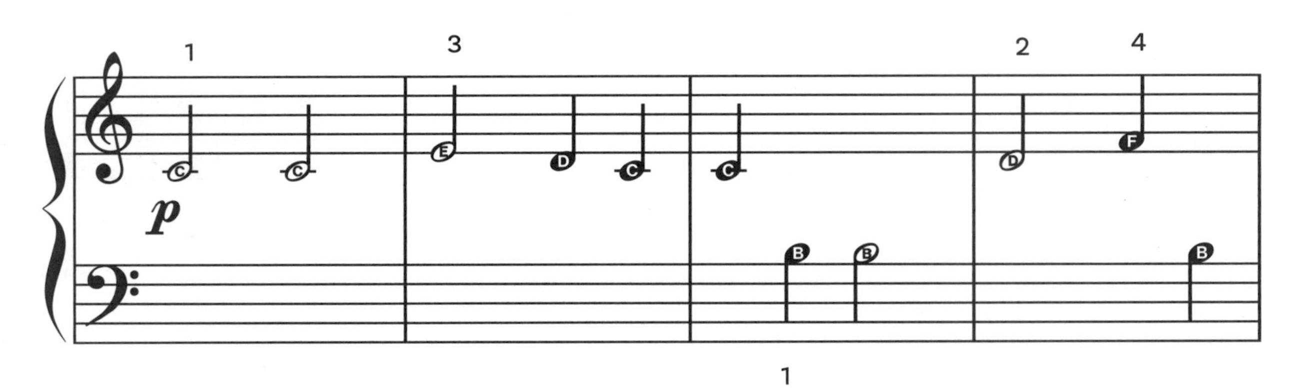
1
3
2
4
p
C C E D C C D F
B B B
1

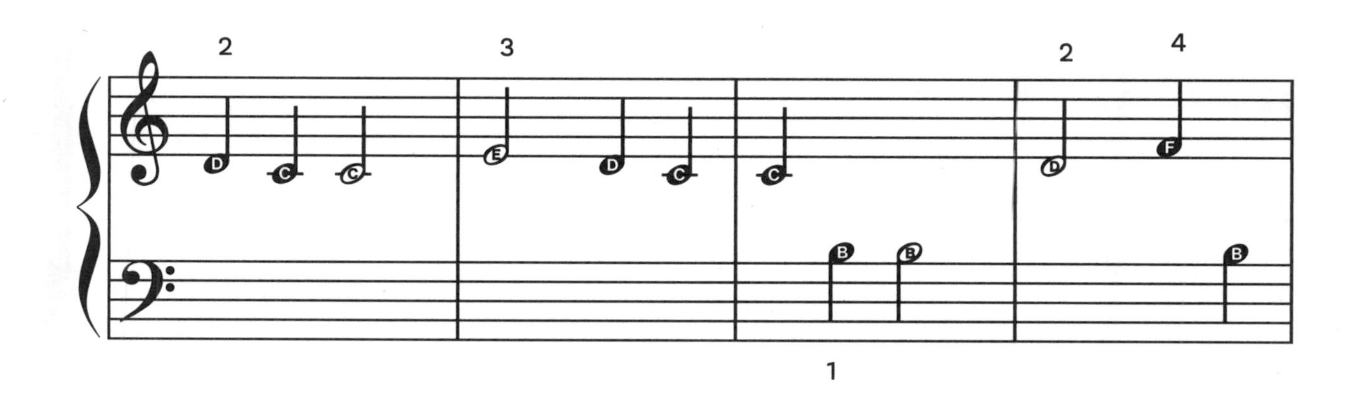
2
D C C
3
E D C
C
2 4
D F
B B
B
1

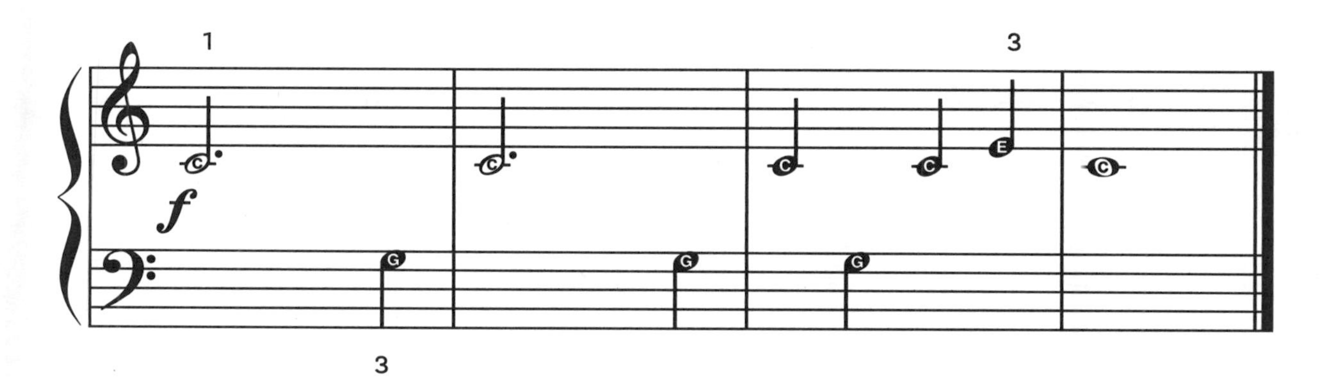
1
C
f
C
C C E
3
C
G G G
3

Ode to Joy

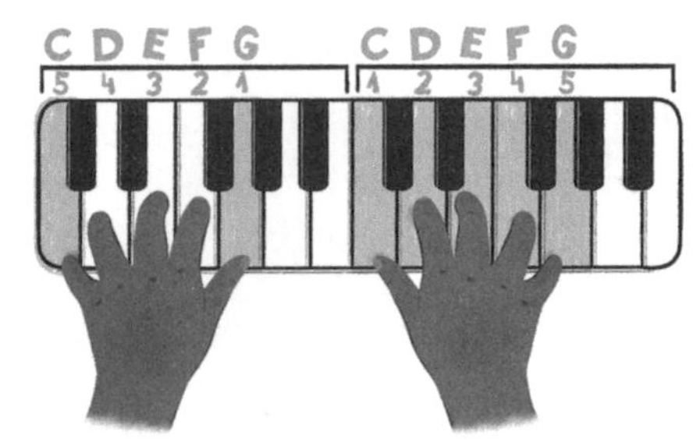
C D E F G
5 4 3 2 1
C D E F G
1 2 3 4 5

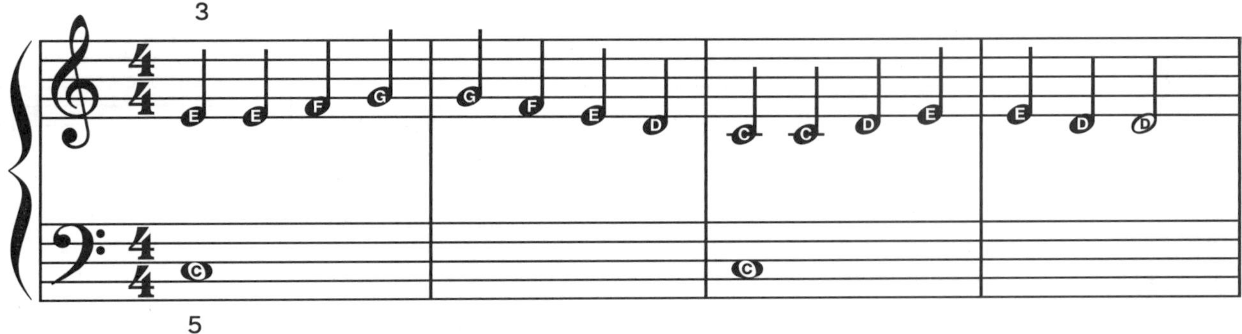
3
E E F G
G F E D
C C D E
E D D
C
C
5

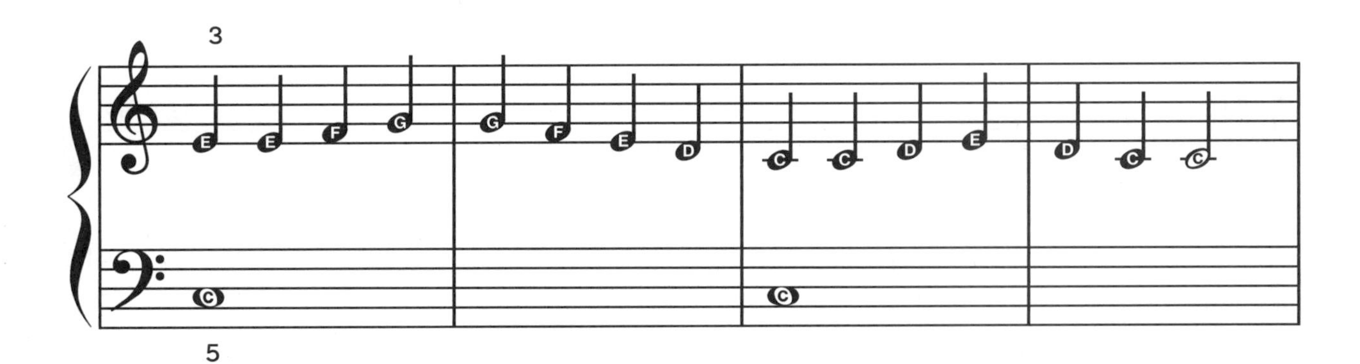
3
E E F G
G F E D
C C D E
D C C
C
C
5

FUN FACT

We often think of classical music as any song written by a composer a long time ago. But only the period between 1760 and 1825 is technically considered the "classical period." One of the most famous classical composers is Ludwig van Beethoven—he wrote this song!

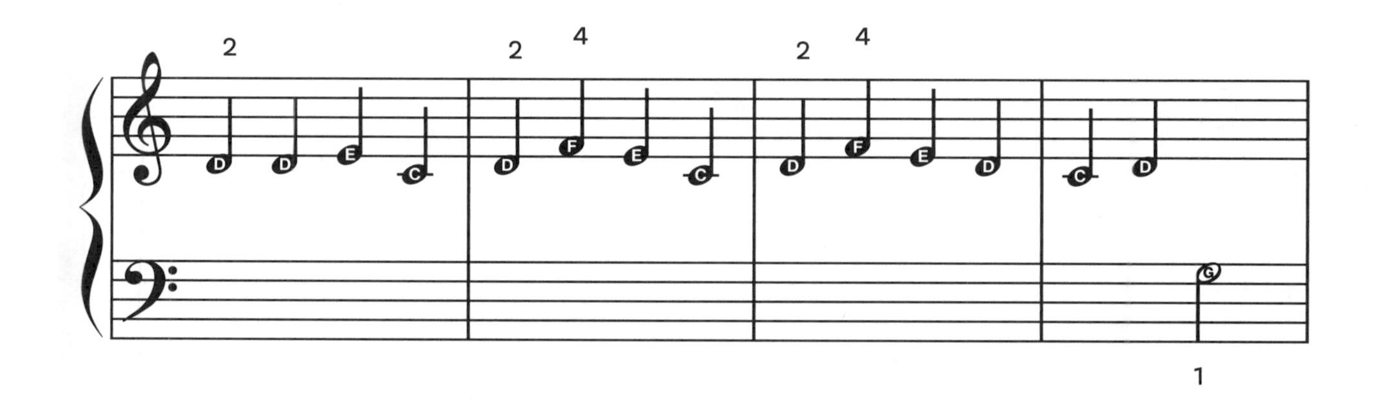
2
2 4
2 4
D D E C
D F E C
D F E D
C D
G
1

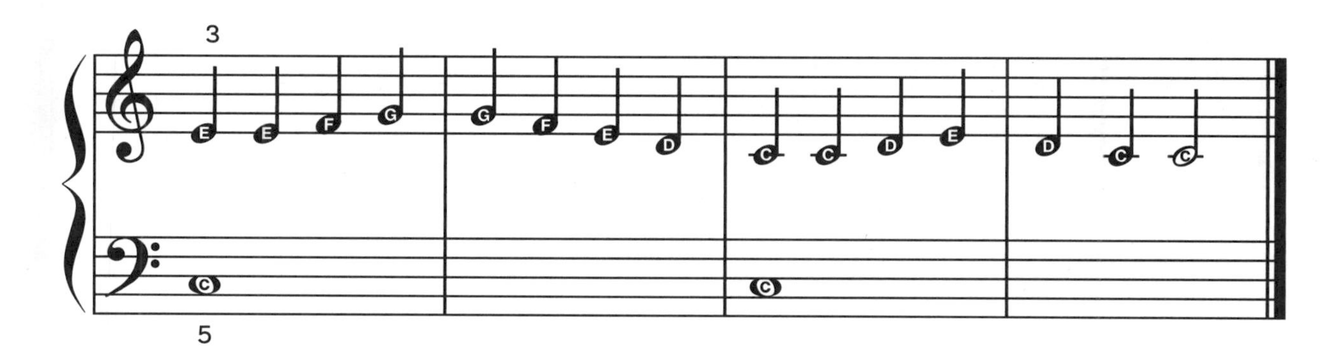
3
E E F G
G F E D
C C D E
D C C
C
C
5

Spring

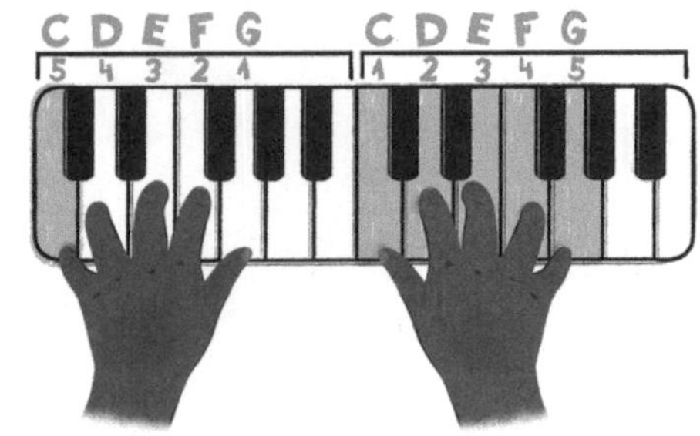
C D E F G
5 4 3 2 1
C D E F G
1 2 3 4 5

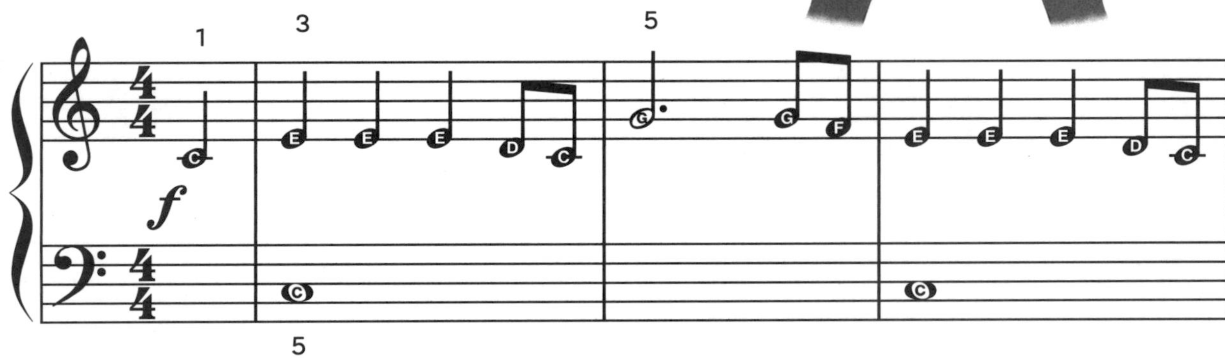
1
3
5
C
E E E D C
G G F
E E E D C
f
C
C
5

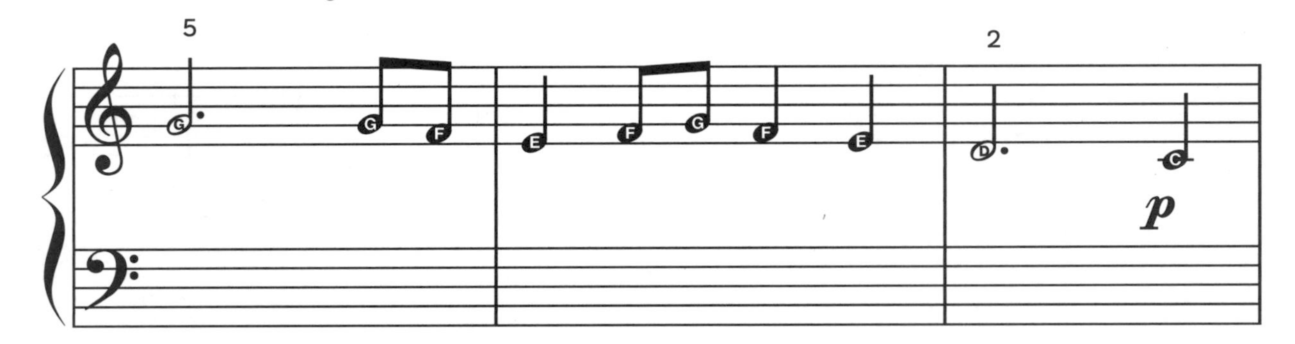
5
G G F
E F G F E
2
D C
p

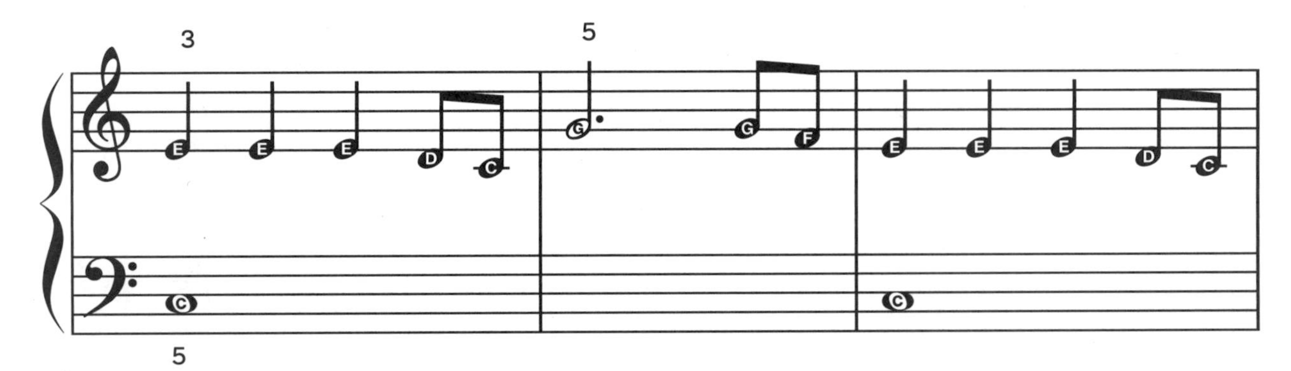
3
E E E D C
5
G G F
E E E D C
C
C
5

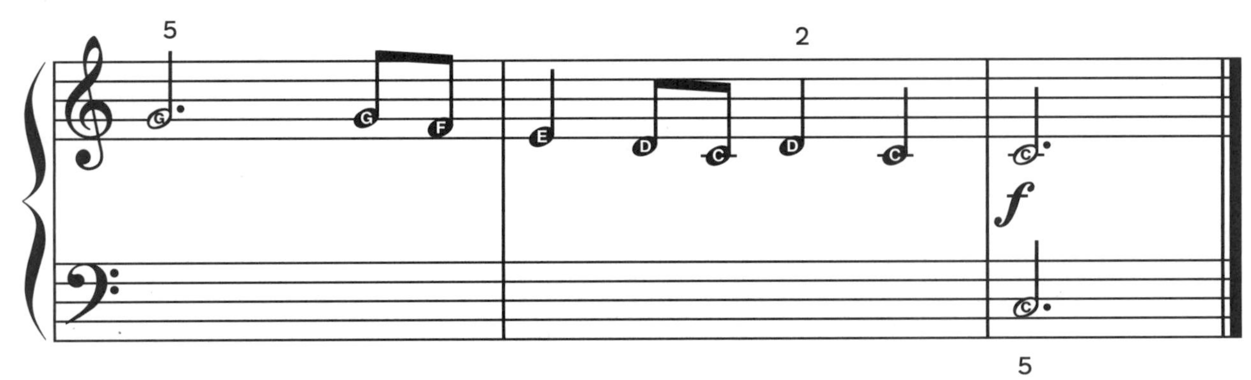
5
G G F
2
E D C D C
C
f
C
5

Finale from *Swan Lake*

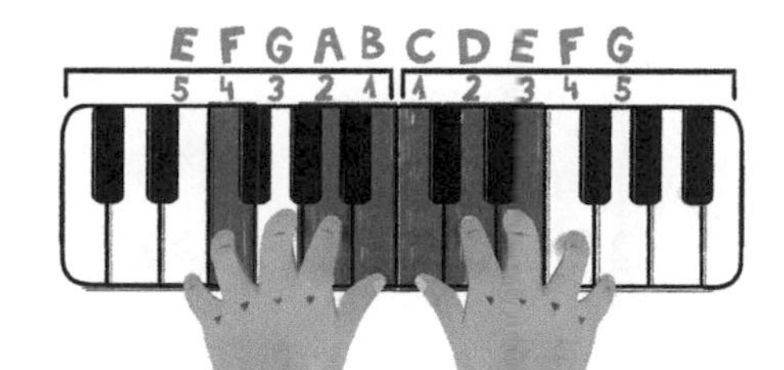
E F G A B C D E F G
5 4 3 2 1 1 2 3 4 5

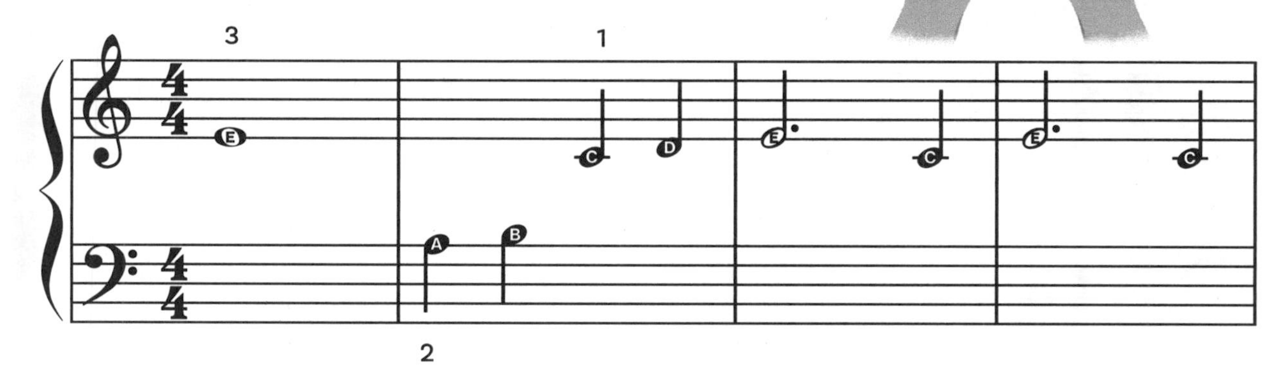

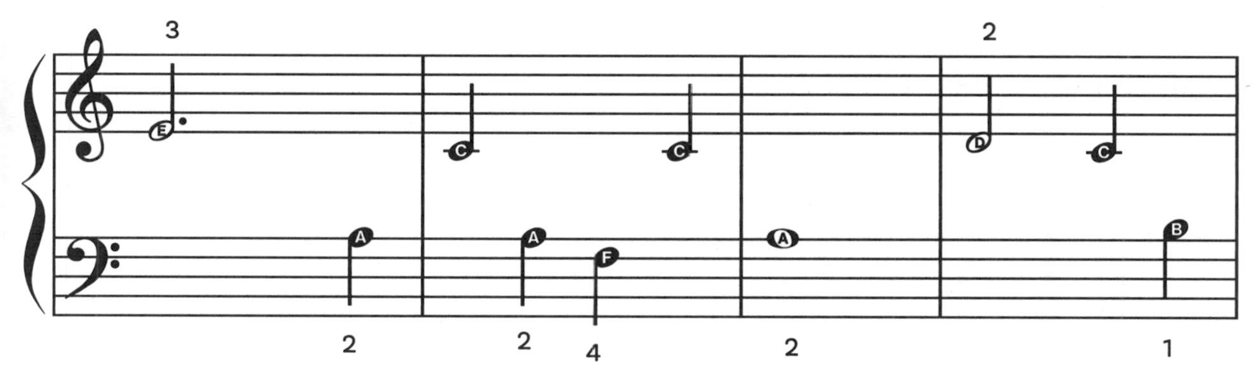

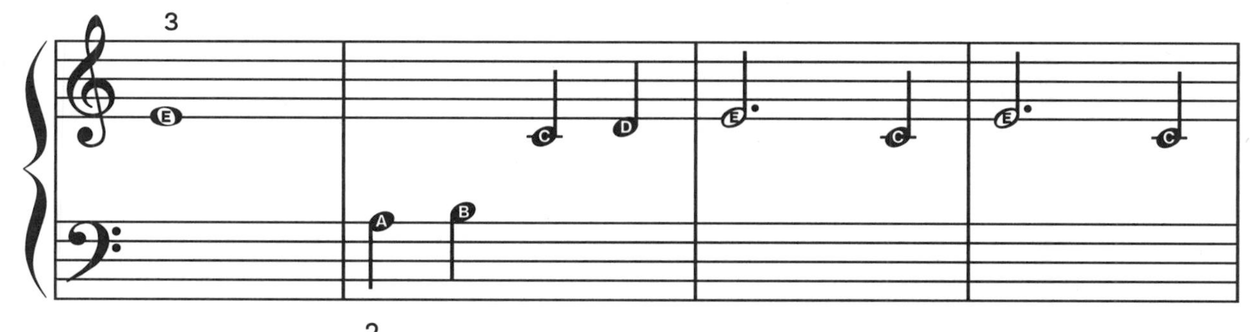

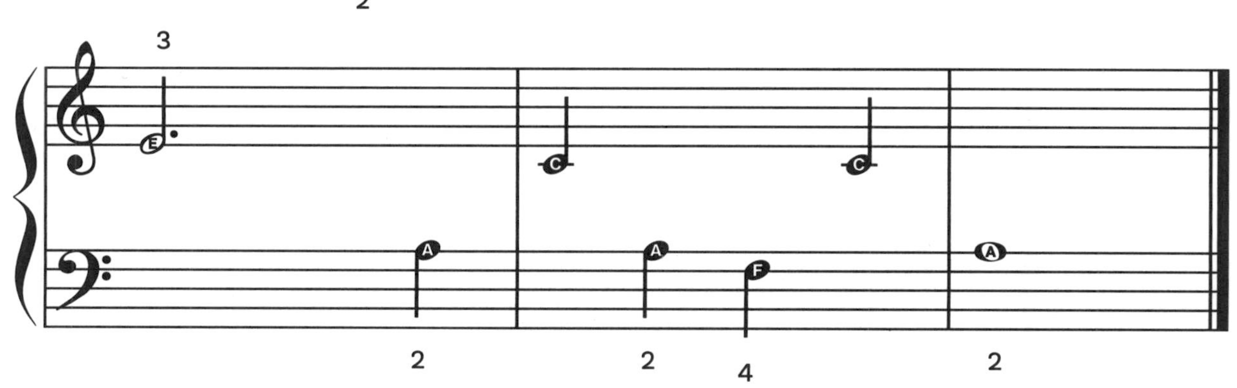

About the Author

Emily Norris has been through her own musical journey!

At age 7, she used her dad's turntable and listened to Bee Gees and Pet Shop Boys records on repeat.

At age 10, she started piano lessons in Tupelo, Mississippi, playing through the whole book in two lessons.

At age 15, she became the music director and piano accompanist for a children's production of *The Wizard of Oz* in Trenton, Tennessee.

At age 22, she taught piano and voice lessons in Clarksville, Tennessee, while also teaching fifth grade.

At age 33, she started her own piano and voice studio in the middle of a global pandemic.

And now, just like G-sharp, she takes young students on their own journeys to learn, love, and enjoy piano!

On days off, she'll go on journeys with her husband on the back of his Honda Shadow motorcycle, still listening to Bee Gees and Pet Shop Boys "records" on repeat.

Parents, you can find Emily online on YouTube (youtube.com/@ebnmusic), Facebook (facebook.com/ebnmusic), Instagram (@ebn.music), and TikTok (@ebnmusic).

About the Illustrator

Malgorzata Detner is a Poland-based illustrator, born in 1989. She currently lives with her family, a cocker spaniel, snails, and two lovely rats in Warsaw. Her love of drawing began at a young age, influenced by her mother's paintings, but grew seriously when she decided to pursue art in middle school. Although Malgorzata initially pursued a career in costume design with an interest in Victorian dresses, her daughter's birth made her return to traditional painting and digital illustration. Influenced by old animation, mysterious, fantastic worlds, animals, and creatures in vibrant colors are what she likes drawing the most. She draws digitally but also likes to incorporate hand-painted textures into her work. Malgorzata loves creating illustrations that remind her of childhood memories.

Parents, you can find Malgorzata on Instagram (@mdetner.illustration) or at malgorzatadetner.com.

To my husband, Caleb,
who is the best supporter in
everything I've done and will do.
You're my everything.
I love you!

Published in the United States by Z Kids, an imprint of Zeitgeist™,
a division of Penguin Random House LLC, New York.
zeitgeistpublishing.com

ISBN: 9780593435793
Ebook ISBN: 9780593690215

Illustrations by Malgorzata Detner
Book design by Katy Brown and Erin Yeung
Author photograph © by Haylee Beth Photography
Illustrator photograph © 2022 by MDetner
Edited by Ada Fung

Printed in the United States of America
1st Printing